Materials Technology for Technicians 2

W. Bolton

Senior Adviser, Technician Education Council

BUTTERWORTHS

TEC

TECHNICIAN SERIES

First published 1981
Reprinted 1985

© Butterworth & Co (Publishers) Ltd, 1981

British Library Cataloguing in Publication Data

Bolton, W.
    Materials technology for technicians 2.
    1. Materials science
    I. Title
    620.1´1   TA403

    ISBN 0 408 01117 3

Typeset by Tunbridge Wells Typesetting Services
Printed and Bound in Great Britain by Page Bros (Norwich) Ltd.

Preface

This book has been written with the following aims:

1. To provide an introduction to the basic properties and structures of metals, both ferrous and non-ferrous, and polymers.
2. To enable some of the reasons for the choice of materials to be appreciated.

The book covers the unit 'Materials Technology 2' of the Technician Education Council, both the general version TEC U78/475 and the version for use with Mechanical and Production Engineering programmes TEC U80/738 being covered. This unit is regarded as giving the basic grounding in materials technology that is required by all mechanical and production engineering technicians. Further units 'Materials Technology 3' and 'Materials Technology 4' develop from this basic understanding and are covered by further books in this series.

Acknowledgements

Thanks are due to the following sources for permission to reproduce illustrations and information used in this book.

Longman Group Ltd
The Royal Society
Hodder and Stoughton Ltd
Edward Arnold Ltd
The Open University
Cassell Ltd
Arthur Lee and Sons Ltd
Sterling Metals Ltd
Ley's Malleable Castings Co. Ltd
McKechnie Metals Ltd
Dynacast International Ltd
Bayer UK Ltd
BCIRA

Contents

1 Structure of metals

Objectives: At the end of this chapter you should be able to:
Explain what is meant by the crystalline state.
Recognise common crystal structures.
Recognise the crystalline nature of metals.
Explain the significance of, and interpret, thermal equilibrium diagrams.
Explain what a solid solution is.
Explain coring.
Explain precipitation hardening.

CRYSTALS

If someone refers to crystals you may well think of material which is geometrically regular in shape, perhaps like the cubes of common salt or sugar. Such crystals have smooth flat faces with the angle between adjoining faces always 90°. Some salt crystals may be small cubes, others large cubes or shapes involving effectively a number of cubes stuck together.

Figure 1.1 shows copper sulphate crystals growing in a drop of solution. Each crystal grows in a regular way, and maintains the same basic shape until it impinges on another crystal or some other obstacle.

The form of crystals and the way in which they grow can be explained if matter is considered to be made up of small particles which are packed together in a regular manner. *Figure 1.2* shows

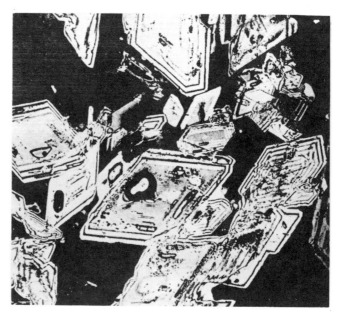

Figure 1.1 Copper sulphate crystals growing from solution. (From Lewis, J., *Physics 11–13,* Longman Group)

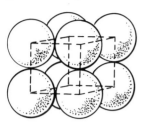

Figure 1.2   A simple cubic structure

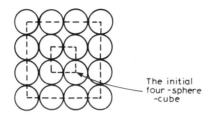

Figure 1.3   A two-dimensional view of the 'growing' four-sphere-cube

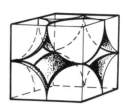

Figure 1.4   The simple cubic structure

how a simple cube can be made by stacking four spheres. The cube can 'grow' if further spheres are added equally to all faces of the four-sphere-cube (*Figure 1.3*). The result is a bigger cube which can be considered to be made up of a larger number of the basic four-sphere-cube.

The dotted lines in *Figure 1.2* enclose what is called the *unit cell*. In this case the unit cell is a cube. The unit cell is the geometric figure which illustrates the grouping of the particles in the solid. This group is repeated many times in space within a crystal, which can be

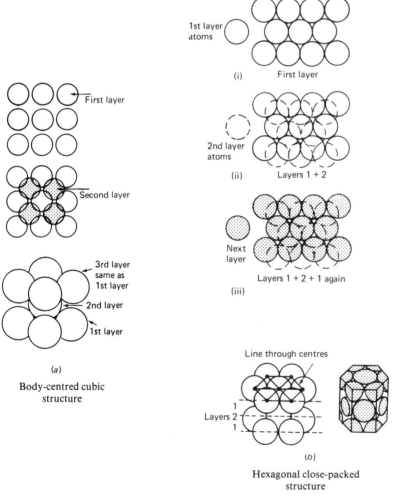

Figure 1.5

considered to be made up, in the case of a *simple cubic crystal,* of a large number of these unit cells stacked together. *Figure 1.4* shows that portion of the stacked spheres that is within the unit cell. The crystal is considered to consist of large numbers of particles arranged in a regular, repetitive pattern, known as the *space lattice,* as in *Figure 1.3.*

It is this regular, repetitive pattern of particles that characterises crystalline material. A solid having no such order in the arrangement of its constituent particles is said to be *amorphous.*

The simple cubic crystal shape is arrived at by stacking spheres in one particular way (*Figure 1.5(a)*). By stacking spheres in different ways, other crystal shapes can be produced (*Figure 1.5(b)* and *(c)*). With the simple cubic unit cell the centres of the spheres lie at the corners of a cube. With the *body-centred cubic* unit cell the cell is slightly more complex than the simple cubic cell in having an extra sphere in the centre of the cell. The *face-centred cubic* cell is another modification of the simple cubic cell, having spheres at the centre of each face of the cube. Another common arrangement is the *hexagonal close-packed* structure.

METALS AS CRYSTALLINE

Metals are crystalline substances. This may seem a strange statement in that metals do not generally seem to look like crystals, with their geometrically regular shapes. If the photograph of the growing copper sulphate crystals, *Figure 1.1,* had been taken a little later, it may well have been difficult to identify the regular shapes of the copper sulphate crystals. Indeed, near the bottom of the photograph the regularity of shape is not apparent. This is where the crystals in their growing have impinged on each other and prevented the individual crystals reaching their geometrically regular shapes. *Figure 1.6* shows a section of a metal. The surface looks much like the surface pattern that would have been produced with the growing copper sulphate crystals if the growth had continued until all the space had been filled.

CHILL CRYSTALS   ⌐COLUMNAR CRYSTALS

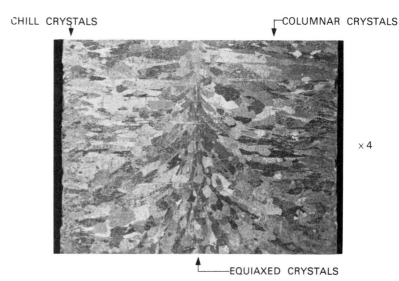

× 4

⌐——EQUIAXED CRYSTALS

Figure 1.6   Cross-section of a small aluminium ingot. (From Monks, H. A. and Rochester, D. C., *Technician Structure and Properties of Metals,* Cassell)

The term *grain* is used to describe the crystals within the metal. A grain is merely a crystal without its geometrical shape and flat faces because its growth was impeded by contact with other crystals. Within a grain the arrangement of particles is just as regular and repetitive as within a crystal with smooth faces. A simple model of a metal with its grains is given by the raft of bubbles on the surface of a liquid (*Figure 1.7*). The bubbles pack together in an orderly and

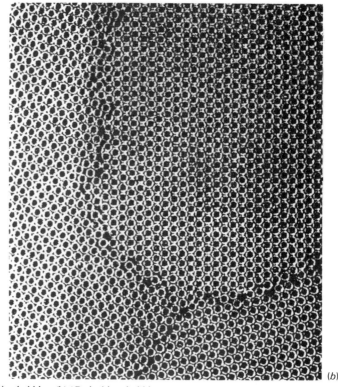

From gas tap

Screw clip

Glass tube drawn out to a fine jet

Soap solution (washing-up liquid in water)

(a)

(b)

Figure 1.7   (a) Simple arrangement for producing bubbles, (b) 'Grains' in a bubble raft. (Courtesy of the Royal Society)

repetitive manner but if 'growth' is started at a number of centres then 'grains' are produced. At the boundaries between the 'grains' the regular pattern breaks down as the pattern changes from the orderly pattern of one 'grain' to that of the next 'grain'.

The grains in the surface of a metal are not generally visible. They can be made visible by careful etching of the surface with a suitable chemical. The chemical preferentially attacks the grain boundaries.

Here are some examples of the different forms of crystal structure adopted by metallic elements.

Body-centred cubic	Face-centred cubic	Hexagonal close-packed
Chromium	Aluminium	Beryllium
Molybdenum	Copper	Cadmium
Niobium	Lead	Magnesium
Tungsten	Nickel	Zinc

GROWTH OF METAL CRYSTALS   Copper sulphate crystals growing in copper sulphate solution are shown in *Figure 1.1*. How do metal crystals grow in liquid metal?

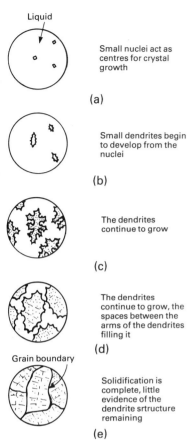

Liquid

Small nuclei act as centres for crystal growth

(a)

Small dendrites begin to develop from the nuclei

(b)

The dendrites continue to grow

(c)

The dendrites continue to grow, the spaces between the arms of the dendrites filling it

(d)

Grain boundary

Solidification is complete, little evidence of the dendrite srtructure remaining

(e)

Figure 1.8 Solidification of a metal

Figure 1.8 shows the various stages that can occur when a metal solidifies. Crystallisation, whether with metals or the copper sulphate, occurs round small nuclei, which may be impurity particles. The initial crystals that form have the shape of the crystal pattern into which the metal normally solidifies, e.g. face-centred cubic in the case of copper. However, as the crystal grows it tends to develop spikes. The shape of the growing crystal thus changes into a 'tree-like' growth called a *dendrite* (*Figure 1.9*). As the dendrite grows so the spaces between the arms of the dendrite fill up. Outward growth of the dendrites cease when the growing arms meet other dendrite arms. Eventually the entire liquid solidifies. When this happens there is little trace of the dendrite structure, only the grains into which the dendrites have grown (see *Figure 1.37*).

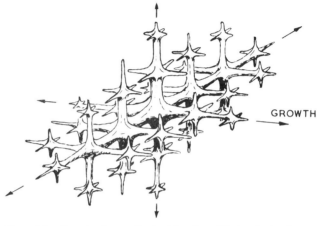

GROWTH

Figure 1.9 Growth of a metallic dendrite (From Higgins, R. A., *Properties of Engineering Materials,* Hodder & Stoughton)

Why do metals tend to grow from the melt as dendrites? Energy is needed to change a solid, at its melting point, to a liquid without any change in temperature occurring; this energy is called *latent heat*. Similarly, when a liquid at the fusion point (i.e. the melting point) changes to a solid, energy has to be removed, no change in temperature occurring during the change of state; this is the latent heat. Thus, when the liquid metal in the immediate vicinity of the metal crystal face solidifies, energy is released which warms up the liquid in front of that advancing crystal face. This slows, or stops, further growth in that direction. The result of this action is that spikes develop as the crystal grows in the directions in which the liquid is coolest. As these warm up in turn, so secondary, and then tertiary, spikes develop as the growth continues in those directions in which the liquid is coolest. This type of growth can be considered in terms of the 'crystal shapes' generated by stacking spheres in an orderly manner (*Figure 1.5*). Instead of stacking the spheres over the entire 'crystal' surface, the spheres are only stacked on parts of that surface (*Figure 1.10*). The result is that although the material is still growing into a crystal with the same unit cell arrangement the growth is not even in all directions; spurs develop.

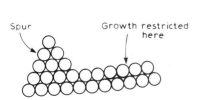

Spur

Growth restricted here

Figure 1.10 Dendritic growth

ALLOYS Brass is an alloy composed of copper and zinc. Bronze is an alloy of copper and tin. An *alloy* is a metallic material consisting of an intimate association of two or more elements. The everyday metallic objects around you will be made almost invariably from alloys rather than the pure metals themselves. Pure metals do not always have the appropriate combination of properties needed; alloys can however be designed to have them.

The coins in your pocket are made of alloys. Coins need to be made of a relatively hard material which does not wear away rapidly, i.e. the coins have to have a 'life' of many years. Coins made of, say, pure copper would be very soft; not only would they suffer considerable wear but they would bend in your pocket.

Coins (British)		Percentage by mass		
	Copper	Tin	Zinc	Nickel
½p, 1p, 2p	97	0.5	2.5	—
5p, 10p, 50p	75	—	—	25

If you put sand in water, the sand does not react with the water but retains its identity, as does the water. The sand in water is said to be a mixture. In a *mixture,* each component retains its own physical structure and properties. Sodium is a very reactive substance, which has to be stored under oil to stop it interacting with the oxygen in the air, and chlorine is a poisonous gas. Yet when these two substances interact, the product, sodium chloride, is eaten by you and me every day. The product is common salt. Sodium chloride is a compound. In a *compound* the components have interacted and the product has none of the properties of its constituents. Alloys are generally mixtures though some of the components in the mixture may interact to give compounds as well.

SOLUBILITY

If you drop a pinch of common salt, sodium chloride, into cold water it will dissolve. The sodium chloride is said to be *soluble* in the cold water. Up to 36 g of sodium chloride can be dissolved in 100 g of cold water; more than that amount will not dissolve. With the 36 g dissolved in 100 g the resulting solution is said to be *saturated.* The *solubility* of sodium chloride in cold water is said to be 36 g per 100 g of water. If 40 g of sodium chloride is put into 100 g of cold water only 36 g of it will dissolve, the remaining 4 g remaining as solid. The solubility of sodium chloride in water depends on the temperature, hot water dissolving more sodium chloride than cold water. The solubility of sodium chloride in water does not however vary to a considerable extent with temperature, only slightly increasing as the temperature increases (*Figure 1.11*).

The solubility of copper sulphate in water does however increase quite significantly with temperature, as *Figure 1.11* shows. *Figure 1.12* shows this solubility variation with temperature plotted in a different way, the temperature axis being vertical rather than horizontal. The reason for this will become apparent later in this chapter when phase diagrams are considered. Suppose we have 40 g of copper sulphate dissolved in 100 g of hot water, say at 90°C. At this temperature the solution will not be saturated. If now the temperature decreases, then at a temperature of just under 60°C the solution will become saturated. Further cooling will result in the excess copper sulphate coming out of solution as a precipitate. At 0°C there will only be about 17 g still in solution.

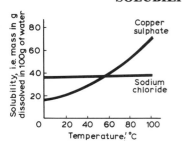

Figure 1.11   The solubility variation with temperature for sodium chloride, common salt, and copper sulphate. The solubility of copper sulphate increases considerably more with temperature than does that of sodium chloride

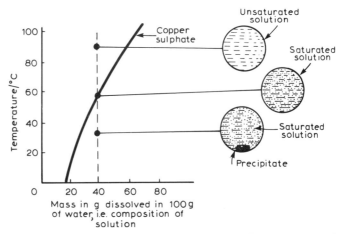

Figure 1.12 An alternative way of plotting in the data given in *Figure 1.11* for the solubility of copper sulphate in water

Some substances, e.g. sand, are not soluble in water. When sand is mixed with water the result is always that none of the sand enters into a solution. The sand is said to be *insoluble* in water.

The terms soluble and insoluble can also be used when we mix two liquids. Oil and water are insoluble in each other; when the two are mixed the result is just a mixture with the oil and the water retaining their separate identities. When alcohol is mixed with water the result is a solution; the water and the alcohol are said to be soluble in each other. It is not possible to identify the water and the alcohol as separate entities in the solution. When two liquid metals are mixed, say copper and nickel, a solution is produced in that it is not possible to identify in the liquid either the copper or the nickel. The copper and the nickel are said to be soluble in each other in the liquid state.

When a salt solution is cooled to such an extent that it solidifies, the resulting solid is a mixture of sodium chloride crystals and ice crystals. In the solid state the salt and the water retain their separate identities; in the solid state they are insoluble in each other. Cadmium and bismuth are soluble in each other in the liquid state but insoluble in each other in the solid state. Copper and nickel, however, are soluble in each other in the solid state. The resulting copper-nickel alloy has, in the solid state, a structure in which it is impossible to distinguish the copper from the nickel. The copper and nickel are thus said to form a *solid solution*.

SOLID SOLUTIONS

Copper forms face-centred cubic crystals (*Figure 1.13a*), as also does nickel (*Figure 1.13b*). When copper and nickel are in solid solution a single face-centred lattice is formed (*Figure 1.13c*). Such a solid solution is said to be *substitutional* in that when, say, nickel is added to the copper, the nickel atoms substitute for the copper atoms in the copper face-centred lattice. This substitution may be ordered, the atoms of the added metal always taking up the same fixed places in the lattice, or it can be disordered. In the disordered solid solution the added atoms appear virtually at random throughout the lattice. Most solid solutions are disordered lattices.

With the copper-nickel solid solution the copper and nickel atoms are virtually the same size. This is necessary for such a solid solution.

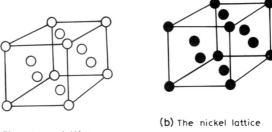

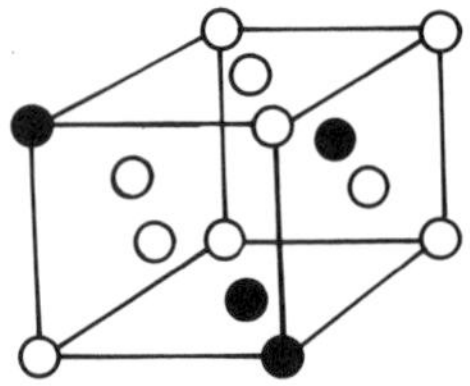

(a) The copper lattice

(b) The nickel lattice

(c) The copper-nickel solid solution lattice

Figure 1.13   (a) The copper lattice, (b) The nickel lattice, (c) The copper-nickel solid solution lattice

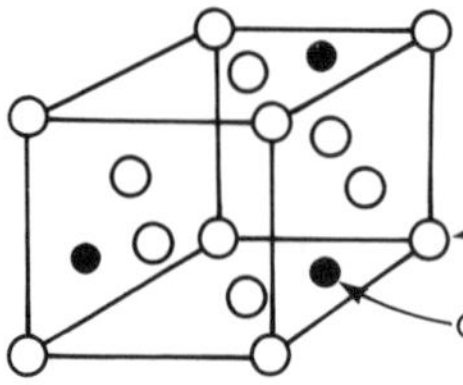

Figure 1.14   An interstitial solid solution. The iron atoms form the face-centred cubic lattice with the carbon atoms fitting in between the iron atoms

Another form of solid solution can, however, occur when the sizes of the two atoms are considerably different. With an *interstitial* solid solution the added atoms are small enough to fit in-between the atoms in the lattice. Carbon can form an interstitial solid solution with the face-centred cubic form of iron. *Figure 1.14* shows the lattice of the solid solution.

ALLOY TYPES

When an alloy is in a liquid state the atoms of the constituents are distributed at random through the liquid. When solidification occurs a number of possibilities exist.

(1) The two components separate out with each in the solid state maintaining its own separate identity and structure. The two components are said to be *insoluble* in each other in the solid state.
(2) The two components remain completely mixed in the solid state. The two components are said to be *soluble* in each other in the solid state, the components forming a solid solution.
(3) On solidifying, the two components may show *limited solubility* in each other.
(4) In solidifying, the elements may combine to form *intermetallic compounds.*

SOLIDIFICATION

When pure water is cooled to 0°C it changes state from liquid to solid, i.e. ice is formed. *Figure 1.15* shows the type of graph that would be produced if the temperature of the water is plotted against time during a temperature change from above 0°C to one below 0°C. Down to 0°C the water only exists in the liquid state. At 0°C solidification starts to occur and while solidification is occurring the temperature remains constant. Energy is still being extracted from the water but there is no change in temperature during this change of

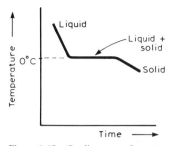

Figure 1.15 Cooling curve for water during solidification

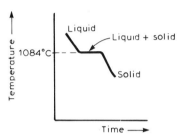

Figure 1.16 Cooling curve for copper during solidification

state. This energy is called *latent heat*. The *specific latent heat of fusion* is defined as the energy taken from, or given to, 1 kg of a substance when it changes from liquid to solid, or solid to liquid, without any change in temperature occurring.

All pure substances show the same type of behaviour as the water when they change state. *Figure 1.16* shows the cooling graph for copper when it changes state from liquid to solid.

During the transition of a pure substance from liquid to solid, or vice versa, the liquid and solid are both in existence. Thus for the water, while the latent heat is being extracted there is both liquid and ice present. Only when all the latent heat has been extracted is there only ice. Similarly with the copper, during the transition from liquid to solid at 1084°C, while the latent heat is being extracted both liquid and solid exist together.

The cooling curves for an alloy do not show a constant temperature occurring during the change of state. *Figure 1.17* shows cooling curves for two copper-nickel alloys. With an alloy, the temperature is not constant during solidification. The temperature range over which this solidification occurs depends on the relative

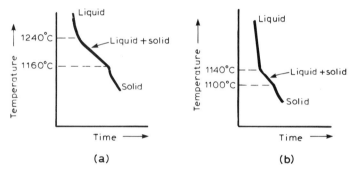

Figure 1.17 Cooling curves for copper-nickel alloys, (a) 70% copper – 30% nickel, (b) 90% copper – 10% nickel

proportions of the elements in the alloy. If the cooling curves are obtained for the entire range of copper-nickel alloys a composite diagram can be produced which shows the effect the relative proportions of the constituents have upon the temperatures at which solidification starts and that at which it is complete. *Figure 1.18* shows such a diagram for copper-nickel alloys.

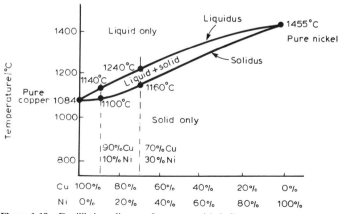

Figure 1.18 Equilibrium diagram for copper-nickel alloys

Thus for pure copper there is a single temperature point of 1084°C, indicating that the transition between liquid and solid takes place at a constant temperature. For 90% copper – 10% nickel the transition between liquid and solid starts at 1140°C and terminates at 1100°C when all the alloy is solid. For 70% copper – 30% nickel the transition between liquid and solid starts at 1240°C and terminates at 1160°C when all the alloy is solid.

The line drawn through the points at which each alloy in the group of alloys ceases to be in the liquid state and starts to solidify is called the *liquidus line.* The line drawn through the points at which each alloy in the group of alloys becomes completely solid is called the *solidus line.* These liquidus and solidus lines indicate the behaviour of each of the alloys in the group during solidification. The diagram in which these lines are shown is called the *thermal equilibrium diagram.*

The thermal equilibrium diagram is constructed from the results of a large number of experiments in which the cooling curves are determined for the whole range of alloys in the group. The diagram provides a forecast of the states that will be present when an alloy of a specific composition is heated or cooled to a specific temperature. The diagrams are obtained from cooling curves produced by very slow cooling of the alloys concerned. They are slow because time is required for equilibrium conditions to obtain at any particular temperature, hence the term thermal equilibrium diagram.

PHASE A *phase* is defined as a region in a material which has the same chemical composition and structure throughout. A piece of pure copper which throughout is the face-centred cubic structure, has but a single phase at that temperature. Molten copper does however represent a different phase in that the arrangement of the atoms in the liquid copper is different from that in the solid copper. A completely homogeneous substance at a particular temperature has only one phase at that temperature. If you take any piece of that homogeneous substance it will show the same composition and structure.

Common salt, in limited quantities, can be dissolved completely in water at given temperature; the salt is said to be soluble in the water. The solution is completely homogeneous throughout, so there is thus just one phase present.

Liquid copper and liquid nickel are completely miscible, as are most liquid metals. The copper-nickel solution is completely homogeneous and thus at the temperature at which the two are liquid there is but one phase present. When the liquid alloy is cooled it solidifies. In the solid state the two metals are completely soluble in each other and so the solid state for this alloy has but one phase. In the case of the 70% copper – 30% nickel alloy (*Figure 1.17a*), the liquid phase exists above 1240°C; between 1240°C and 1160°C there are two phases when both liquid and solid are present. Below 1160°C the copper-nickel alloy exists in just one phase as the two metals are completely soluble in each other and give a solid solution. The thermal equilibrium diagram given in *Figure 1.18* for the range of copper-nickel alloys is thus a diagram showing the phase or phases present at any particular temperature for any composition of copper-nickel alloy.

Example

What phase is present at 1200°C for the 40% copper – 60% nickel alloy?

Using *Figure 1.18* the line for the 40% copper – 60% nickel alloy meets the solidus at a temperature greater than 1200°C. Hence the phase present at that temperature is the solid.

Example

A copper-nickel alloy is required to be liquid at temperatures down to 1140°C. What is the alloy with the highest percentage nickel for which this can occur?

Using *Figure 1.18* the liquidus at 1140°C gives an alloy with 90% copper and 10% nickel. This is the alloy with the highest percentage nickel which is just liquid at 1140°C.

EQUILIBRIUM DIAGRAMS AND SOLUBILITY

The equilibrium diagram for the copper-nickel alloy is typical of that given when two components are soluble in each other, both in the liquid and solid states. *Figure 1.19* shows the general form of such an equilibrium diagram.

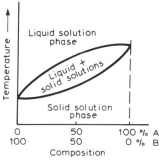

Figure 1.19 Equilibrium diagram for two metals that are completely soluble in each other in the liquid and solid states

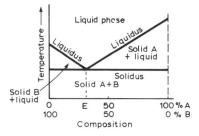

Figure 1.20 Equilibrium diagram for two metals that are completely soluble in each other in the liquid state and completely insoluble in each other in the solid state

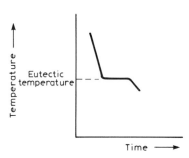

Figure 1.21 Cooling curve for the eutectic composition

Figure 1.20 shows the type of equilibrium diagrams produced when the two alloy components are completely soluble in each other in the liquid state but completely insoluble in each other in the solid state. The solid alloy shows a mixture of crystals of the two metals concerned. Each of the two metals in the solid alloy retains its independent identity. At one particular composition, called the *eutectic composition* (marked as E in *Figure 1.20*), the temperature at which solidification starts to occur is a minimum. At this temperature, called the *eutectic temperature,* the liquid changes to the solid state without any change in temperature (*Figure 1.21*). The solidification at the eutectic temperature, for the eutectic composition, has both the metals simultaneously coming out of the liquid. Both metals crystallise together. The resulting structure, known as the *eutectic structure,* is generally a laminar structure with layers of metal A alternating with layers of metal B (*Figure 1.22*).

The properties of the eutectic can be summarised as:

(1) Solidification takes place at a single fixed temperature:
(2) The solidification takes place at the lowest temperature in that group of alloys.

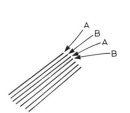

Figure 1.22   The laminar structure of the eutectic

(3) The composition of the eutectic composition is a constant for that group of alloys.

(4) It is a mixture, for an alloy made up from just two metals, of the two phases.

(5) The solidified eutectic structure is generally a laminar structure.

Figure 1.23 illustrates the sequence of events that occur when the 80% A–20% B liquid alloy is cooled. In the liquid state both metals are completely soluble in each other and the liquid alloy is thus

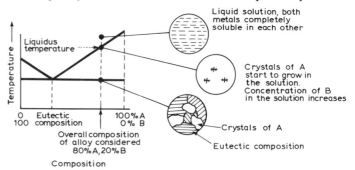

Figure 1.23

completely homogeneous. When the liquid alloy is cooled to the liquidus temperature, crystals of metal A start to grow. This means that as metal A is being withdrawn from the liquid, the composition of the liquid must change to a lower concentration of A and a higher concentration of B. As the cooling proceeds and the crystals of A continue to grow so the liquid further decreases in concentration of A and increases in concentration of B. This continues until the concentrations in the liquid reach that of the eutectic composition. When this happens solidification of the liquid gives the eutectic structure. The resulting alloy has therefore crystals of A embedded in a structure having the composition and structure of the eutectic. *Figure 1.24* shows the cooling curve for this sequence of events.

Apart from an alloy having the eutectic composition and structure when the alloy is entirely of eutectic composition, all the other alloy compositions in the alloy group shows crystals of either metal A or B embedded in eutectic structure material (*Figure 1.25*). Thus for two metals that are completely insoluble in each other in the solid state:

(1) The structure prior to the eutectic composition is of crystals of B in eutectic composition and structure material.

(2) At the eutectic structure the material is entirely eutectic composition and structure.

(3) The structure after the eutectic composition is of crystals of A in eutectic composition and structure material.

Example
Figure 1.26 shows the equilibrium diagram for alloys of bismuth and cadmium. (a) What is the composition of the eutectic? (b) What will be the structure of a solid 80% cadmium – 20% bismuth alloy? (c) What will be the structure of a solid 20% cadmium – 80% bismuth alloy?

(a) The eutectic composition is 40% cadmium – 60% bismuth.
(b) Crystals of cadmium in eutectic.
(c) Crystals of bismuth in eutectic.

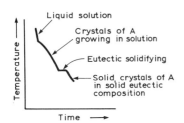

Figure 1.24   Cooling curve for an 80% A – 20% B alloy

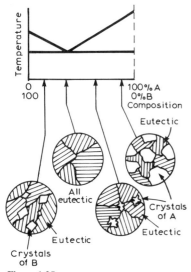

Figure 1.25

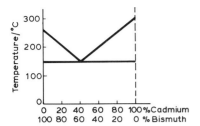

Figure 1.26 Equilibrium diagram for cadmium-bismuth alloys

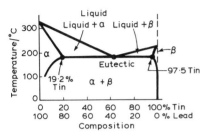

Figure 1.27 Equilibrium diagram for lead-tin alloys

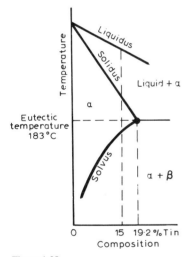

Figure 1.28

Many metals are neither completely soluble in each other in the solid state nor completely insoluble; each of the metals is soluble in the other to some limited extent. Lead-tin alloys are of this type. *Figure 1.27* shows the equilibrium diagram for lead-tin alloys. The solidus line is that line, starting at 0% tin – 100% lead, between the (liquid + α) and the α areas, between the (liquid + α) and the ($\alpha + \beta$) areas, between the (liquid + β) and the ($\alpha + \beta$) areas, and between the (liquid + β) and the β areas. The α, the β, and the ($\alpha + \beta$) areas all represent solid forms of the alloy. The transition across the line between α and ($\alpha + \beta$) is thus a transition from one solid form to another solid form. Such a line is called the *solvus*. *Figure 1.28* shows the early part of *Figure 1.27* and the liquidus, solidus and solvus lines.

Consider an alloy with the composition 15% tin – 85% lead (*Figure 1.28*). When this cools from the liquid state, where both metals are soluble in each other, to a temperature below the liquidus then crystals of the α phase start to grow. The α phase is a solid solution. Solidification becomes complete when the temperature has fallen to that of the solidus. At that point the solid consists entirely of crystals of the α phase. This solid solution consists of 15% tin completely soluble in the 85% lead at the temperature concerned. Further cooling results in no further change in the crystalline structure until the temperature has fallen to that of the solvus. At this temperature the solid solution is saturated with tin. Cooling below this temperature results in tin coming out of solution in another solid solution β. The more the alloy is cooled the greater the amount of tin that comes out of solution, until at room temperature most of the tin has come out of the solid solution. The result is largely solid solution crystals, the α phase, having a low concentration of tin in lead, mixed with small solid solution crystals, the β phase, having a high concentration of tin in lead.

At the eutectic temperature the maximum amount of tin that can be dissolved in lead in the solid state is 19.2% (see *Figure 1.28*). Similarly the maximum amount of lead that can be dissolved in tin, at the eutectic temperature, is 2.5%.

The eutectic composition is 61.9% tin – 38.1% lead. When an alloy with this composition is cooled to the eutectic temperature the behaviour is the same as when cooling to the eutectic occurred for the two metals insoluble in each other in solid state (*Figure 1.23*) except that instead of pure metals separating out to give a laminar mixture of the metal crystals there is a laminar mixture of crystals of the two solid solutions α and β. The α phase has the composition of 19.2% tin – 80.8% lead, the β phase has the composition 97.5% tin – 2.5% lead. Cooling below the eutectic temperature results in the α solid solution giving up tin, due to the decreasing solubility of the tin in the lead, and the β solid solution giving up lead, due to the decreasing solubility of the lead in the tin. The result at room temperature is a structure having a mixture of alpha and beta solid solution, the alpha solid solution having a high concentration of lead and the beta a high concentration of tin.

For alloys having a composition with between 19.2% and 61.9% tin, cooling from the liquid results in crystals of the α phase separating out when the temperature falls below that of the liquidus. When the temperature reaches that of the solidus, solidification is complete and the structure is that of crystals of the α solid solution in eutectic structure material. Further cooling results in α solid solution loosing tin. The eutectic mixture has the α part of it loosing tin and the β part loosing lead. The result at room temperature is a structure having the α solid solution crystals with a high concentration of lead and very little tin and some β precipitate, and the eutectic structure a mixture of α with high lead concentration and β with high tin concentration.

For alloys having a composition with between 61.9% and 97.5% tin, cooling from the liquid results in crystals of the β phase separating out when the temperature falls below that of the liquidus. Otherwise the events are the same as those occurring for compositions between 19.2% and 61.9% tin. The result at room temperature is a structure having the β solid solution crystals with a high concentration of tin and very little lead and some α precipitate, and the eutectic structure a mixture of α with high lead concentration and β with high tin concentration.

For alloys having a composition with more than 97.5% tin present, crystals of the β phase begin to grow when the temperature falls below that of the liquidus. When the temperature falls to that of the solidus solidification becomes complete and the solid consists entirely of β solid solution crystals. Further cooling results in no further change in the structure until the temperature reaches that of the solvus. At this temperature the solid solution is saturated with lead and cooling below this temperature results in the lead coming out of solution. The result at room temperature, when most of the lead has come out of the solid solution, is β phase crystals having a high concentration of tin mixed with α phase crystals with a high concentration of lead.

Figure 1.29 shows the types of structure that might be expected at room temperature for lead-tin alloys of different compositions.

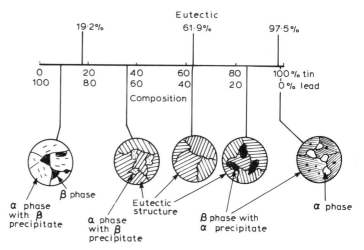

Figure 1.29   Lead-tin alloys

Example

Figure 1.30 shows the equilibrium diagram for silver-copper alloys. What will be the structure at room temperature of (a) a 5% copper –95% silver alloy, (b) a 28.5% copper – 81.5% silver alloy, (c) a 70% copper – 30% silver alloy, (d) a 35% copper – 65% silver alloy?

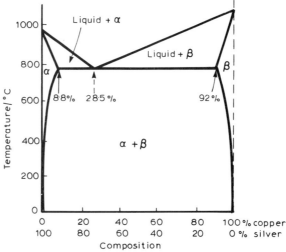

Figure 1.30 Equilibrium diagram for copper-silver alloys

(a) Solid solution α phase crystals, largely silver, mixed with small solid solution β crystals, largely copper.

(b) This is the eutectic composition and so the entire alloy will be eutectic structure.

(c) Solid solution β phase crystals, largely copper, mixed with eutectic structure material.

(d) Solid solution β phase crystals, largely copper, mixed with eutectic structure material. The β phase crystals are smaller than in (c). *Figure 1.31* shows the structure.

In many cases of alloys, the phase diagrams are more complex than those already considered in this chapter. The complexity occurs because of the formation of further phases. These can be due to the formation of intermetallic compounds. *Figure 1.32* shows the

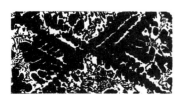

Figure 1.31 Copper dendrites in copper-silver eutectic. (From Rollason, E. C., *Metallurgy for Engineers,* Edward Arnold)

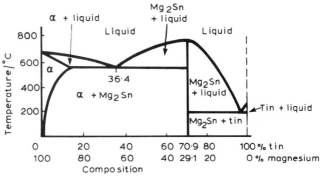

Figure 1.32 Magnesium-tin equilibrium diagram

magnesium-tin thermal equilibrium diagram. This can be considered to be essentially two equilibrium diagrams stuck together, the dividing line between the two occurring at 29.1% magnesium –70.9% tin. This is the composition of the intermetallic compound between magnesium and tin (Mg_2Sn).

Because it is possible to consider each part of the magnesium-tin thermal equilibrium diagram separately, the diagram up to 70.9% tin can be considered as being that due to two materials partially soluble in each other in the solid state, the substances between magnesium and the intermetallic compound. Between 70.9% and 100% tin the equilibrium diagram is like that of two materials insoluble in each other in the solid state, the materials being tin and the intermetallic compound. Thus an alloy of composition 20% tin – 80% magnesium will, at room temperature, have α phase and eutectic structure, this being a mixture of α phase and intermetallic compound. An alloy of 36.4% tin – 63.6% magnesium will have purely the eutectic structure, a mixture of α phase and intermetallic compound. An alloy of 80% tin – 20% magnesium will consist of intermetallic compound crystals in a eutectic composed of a mixture of intermetallic compound and tin.

Other forms of thermal equilibrium diagrams are produced when, during the cooling process from the liquid, a reaction occurs between the solid that is first produced and the liquid in which it is forming. The reaction is called a *peritectic reaction*. *Figure 1.33* shows part of the thermal equilibrium diagram for copper-zinc alloys, peritectic reactions occurring during the solidification of such alloys.

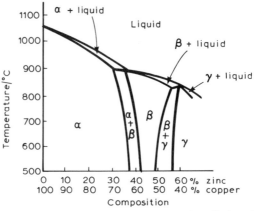

Figure 1.33   Part of the copper-zinc thermal equilibrium diagram

NON-EQUILIBRIUM CONDITIONS

The term 'equilibrium' has been used in connection with the thermal equilibrium diagrams used to predict the structure of alloys. This has been considered to mean a very slow cooling of the alloy from the liquid state. Why should the cooling be slow?

Consider the solidification of a copper-nickel alloy, e.g. 70% copper – 30% nickel. *Figure 1.34* shows the relevant part of the thermal equilibrium diagram. When the liquid copper-nickel alloy cools to the liquidus temperature, small dendrites of copper–nickel solid solution form. Each dendrite will have the composition of 53% copper – 47% nickel. This is the composition of the solid that can be in equilibrium with the liquid at the temperature concerned, the composition being obtained from the thermal equilibrium diagram by drawing a constant temperature line at this temperature and finding the intersection of the line with the solidus. As the overall composition of the liquid plus solid is 70% copper – 30% nickel, the dendrites, in having a greater percentage of nickel, mean that the remaining liquid must have a lower concentration of nickel than 30%.

As the alloy cools further, so the dendrites grow. At 1200°C the expected composition of the solid material would be 62% copper

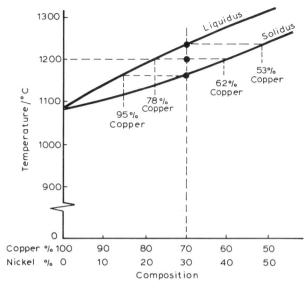

Figure 1.34 Thermal equilibrium diagram for copper rich alloys of copper-nickel

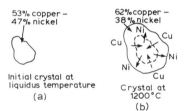

Initial crystal at
liquidus temperature
(a)

Crystal at
1200°C
(b)

Figure 1.35 Diffusion during crystal growth, (a) Initial crystal at the liquidus temperature, (b) Crystal at 1200°C

–38% nickel, the liquid having the composition 78% copper – 32% nickel. So the percentage of copper has increased from the 53% at the liquidus temperature, while the percentage of nickel has decreased. If the dendrite is to have a constant composition, movement of atoms within the solid will have to occur. The term *diffusion* is used for the migration of atoms. The nickel atoms will have to move outwards from the initial dendrite core and copper atoms will have to move inwards to the core; *Figure 1.35* illustrates this process. This diffusion takes time, in fact the process of diffusion in a solid is very slow.

As the alloy cools further so the expected composition of the solid material changes until at the solidus temperature the composition becomes 70% copper – 30% nickel, the last drop of the liquid having the composition 95% copper – 5% nickel. For the solid to have this uniform composition there must have been a diffusion of copper atoms inwards to the core of the dendrite and a corresponding movement of nickel atoms outwards. For this to happen the entire process of cooling from the liquid must take place very slowly. This is what is meant by equilibrium conditions.

In the normal cooling of an alloy, in perhaps the production of a casting, the time taken for the transition from liquid to solid is relatively short and inadequate for sufficient diffusion to have occurred for constant composition solid to be achieved.

The outcome is that the earlier parts of the crystal growth have a higher percentage of nickel than the later growth parts. The earlier growth parts have, however, a lower percentage of copper than the later growth parts (*Figure 1.36*). This effect is called *coring,* the more rapid the cooling from the liquid the more pronounced the coring, i.e. the greater the difference in composition between the earlier and later growth parts of crystals. *Figure 1.37* shows the cored structure of the 70% copper – 30% nickel alloy. The photograph shows the etched surface of the alloy; as the amount of etching that takes place is determined by the composition of the metal, the earlier and later parts of the dendritic growth are etched to different degrees and so

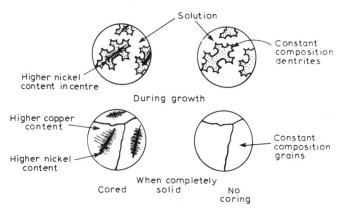

Figure 1.36   Coring with a 70% copper – 30% nickel alloy

show on the photograph as different degrees of light and dark. The effect of this is to show clearly the earlier parts of the dendritic growth within a crystal grain.

Coring can be eliminated after an alloy has solidified by heating it to a temperature just below that of the solidus and then holding it at that temperature for a sufficiently long time to allow diffusion to occur and a uniform composition be achieved.

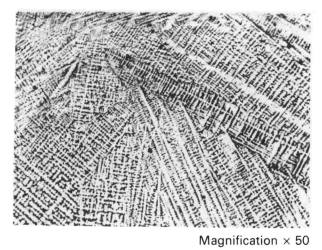

Magnification × 50

Figure 1.37   The cored structure in a 70% copper – 30% nickel alloy. (Courtesy of the Open University)

PRECIPITATION   If a solution of sodium chloride in water is cooled sufficiently, sodium chloride precipitates out of the solution. This occurs because the solubility of sodium chloride in water decreases as the temperature decreases. Thus very hot water may contain 37 g per 100 g of water. Cold water is saturated with about 36 g. When the hot solution cools down the surplus salt is precipitated out of the solution. Similar events can occur with solid solutions.

Figure 1.38 shows part of the copper-silver thermal equilibrium diagram (i.e. part of *Figure 1.30*). When the 5% copper – 95% silver alloy is cooled from the liquid state to 800°C a solid solution is produced. At this temperature the solid solution is not saturated but, cooling to the solvus temperature makes the solid solution saturated.

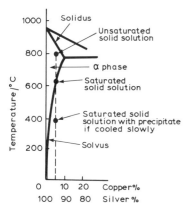

Figure 1.38 Copper-silver thermal equilibrium diagram

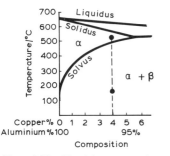

Figure 1.39 Aluminium-copper thermal equilibrium diagram

If the cooling is continued, slowly, a precipitate occurs. The result at room temperature is a solid solution in which a coarse precipitate occurs.

The above discussion assumes that the cooling occurs very slowly. The formation of a precipitate requires the grouping together of atoms. This requires atoms to diffuse through the solid solution. Diffusion is a slow process, if the solid solution is cooled rapidly from 800°C, i.e. quenched, the precipitation may not occur. The solution becomes *supersaturated,* i.e. it contains more of the α phase than the equilibrium diagram predicts. The result of this rapid cooling is a solid solution, the α phase, at room temperature.

The supersaturated solid solution may be retained in this form at room temperature, but the situation is not very stable and a very fine precipitation may occur with the elapse of time. This precipitation may be increased if the solid is heated for some time (the temperature being significantly below the solvus temperature). The precipitate tends to be very minute particles dispersed throughout the solid. Such a fine dispersion gives a much stronger and harder alloy than when the alloy is cooled slowly from the α solid solution. This hardening process is called *precipitation hardening.* The term *natural ageing* is sometimes used for the hardening process that occurs due to precipitation at room temperature and the term *artificial ageing* when the precipitation occurs as a result of heating.

Figure 1.39 shows part of the thermal equilibrium diagram for aluminium-copper alloys. If the alloy with 4% copper is heated to about 500°C and held at that temperature for a while, diffusion will occur and a homogeneous α solid solution will form. If the alloy is then quenched to about room temperature supersaturation occurs. This quenched alloy is relatively soft. If now the alloy is heated to a temperature of about 165°C and held at this temperature for about ten hours a fine precipitate is formed. *Figure 1.40* shows the effects on the alloy structure and properties of these processes. The effect is to give an alloy with a higher tensile strength and harder.

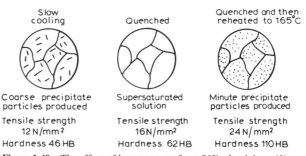

Figure 1.40 The effect of heat tretment for a 96% aluminium–4% copper alloy

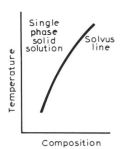

Figure 1.41 The required form of solvus line for precipitation hardening

Not all alloys can be treated in this way. Precipitation hardening can only occur, in a two-metal alloy, if one of the alloying elements has a high solubility at high temperatures and a low solubility at low temperatures, i.e. the solubility decreases as the temperature decreases. This means that the solvus line must slope as shown in *Figure 1.41.* Also the structure of the alloy at temperatures above the solvus line must be a single phase solid solution. The alloy systems that have some alloy compositions that can be treated in this way are mainly non-ferrous, e.g. copper-aluminium and magnesium-aluminium.

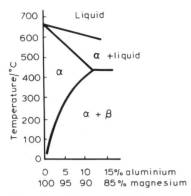

Figure 1.42   Thermal equilibrium diagram for aluminium-magnesium alloys

Example

Figure 1.42 shows part of the thermal equilibrium diagram for magnesium-aluminium alloys. A 10% aluminium – 90% magnesium alloy is heated to 450°C and held at that temperature for a while before being cooled rapidly to room temperature by quenching. What type of structure would you expect the alloy to have?

That of a supersaturated solid solution. No precipitate would be anticipated.

Example

Which of the following alloys would you anticipate could be precipitation hardened?

(a) 95% magnesium – 5% aluminium alloy (see *Figure 1.42*).
(b) 85% magnesium – 15% aluminium alloy (see *Figure 1.42*).
(c) 20% cadmium – 80% bismuth alloy (see *Figure 1.26*).
(d) 5% copper – 95% silver alloy (see *Figure 1.30*).
(e) 95% copper – 5% silver alloy (see *Figure 1.30*).

(a), (d) and (e) would be expected to be capable of being precipitation hardened on the basis of the slope of the solvus in their thermal equilibrium diagrams. (b) Has a composition that has too much aluminium to permit precipitation hardening. Cadmium-bismuth alloys (c) cannot be precipitation hardened.

PROBLEMS

1. Describe the structure of the unit cells of simple cubic, body-centred cubic, face-centred cubic and close-packed hexagonal crystals.

2. Describe the growth of a grain in a metal from the nuclear crystal through dendritic growth.

3. What is the difference between a mixture of two substances and a compound of the two?

4. Explain what is meant by the terms soluble, insoluble, liquid solution, solid solution.

5. Sketch the form of cooling curve you would expect when a sample of pure iron is cooled from the liquid to solid state.

6. *Figure 1.43* shows the cooling curves for copper-nickel alloys. Use these to plot the copper-nickel thermal equilibrium diagram.

7. Use either *Figure 1.18* or your answer to Problem 6 to determine the liquidus and solidus temperatures for a 50% copper – 50% nickel alloy.

8. Germanium and silicon are completely soluble in each other in both the liquid and solid states. Plot the thermal equilibrium diagram for germanium-silicon alloys from the following data.

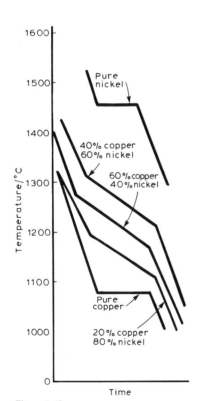

Figure 1.43

Alloy		Liquidus temperature	Solidus temperature
Germanium %	Silicon %	/°C	/°C
100	0	958	958
80	20	1115	990
60	40	1227	1050
40	60	1315	1126
20	80	1370	1230
0	100	1430	1430

9. Explain what is meant by the liquidus, solidus and solvus lines on a thermal equilibrium diagram.

10. Describe the form of the thermal equilibrium diagrams that would be expected for alloys of two metals that are completely soluble in each other in the liquid state but in the solid state are (a) soluble, (b) completely insoluble, (c) partially soluble in each other.

11. The lead-tin thermal equilibrium diagram is given in *Figure 1.27*
(a) What is the composition of the eutectic?
(b) What is the eutectic temperature?
(c) What will be the expected structure of a solid 40% tin – 60% copper alloy?
(d) What will be the expected structure of a solid 10% tin – 90% copper alloy?
(e) What will be the expected structure of a solid 90% – 10% copper alloy?

12. The bismuth-cadmium thermal equilibrium diagram is given in *Figure 1.26.*
(a) At what temperature would you expect a 20% cadmium – 80% bismuth alloy to begin to solidify from the liquid?
(b) What is the composition of the alloy which is liquid at the lowest temperature?

13. What are the liquidus and solidus temperatures for a 80% copper – 20% silver alloy (see *Figure 1.30*)?

14. For (a) a 40% tin – 60% lead alloy, (b) an 80% tin – 20% lead alloy, what are the phases present at tempertaures of (i) 250°C, (ii) 200°C? (See *Figure 1.27*).

15. Explain what is meant by coring and the conditions under which it occurs.

16. Explain how precipitation hardening is produced.

17. The relevant part of the aluminium-copper thermal equilibrium diagram is given in *Figure 1.39*. What type of microstructure would you expect for a 2% copper – 98% aluminium alloy after it has been heated to 500°C, held at that temperature for a while, and then cooled (a) very slowly to room temperature, (b) very rapidly to room temperature?

18. Use the information given in the thermal equilibrium diagram for lead-tin alloys, *Figure 1.27,* for this question.
(a) Sketch the cooling curves for liquid to solid transitions for (i) 20% tin – 80% lead, (ii) 40% tin – 60% lead, (iii) 60% tin – 40% lead, (iv) 80% tin – 20% tin alloys.
(b) Solder used for electrical work in the making of joints between wires has about 67% lead – 33% tin. Why do you consider this alloy composition is chosen?
(c) What is the lowest temperature at which lead-tin alloys are liquid?

2 Carbon steels

Objectives: At the end of this chapter you should be able to:
Describe the iron-carbon system, explaining the effects of percentage carbon on the properties of carbon steel.
Explain the basic heat treatment processes in terms of the iron-carbon thermal equilibrium diagram.
Describe the processes used for the surface hardening of carbon steels.
Select carbon steels and heat treatment processes for specific applications.

IRON ALLOYS

Pure iron is a relatively soft material and is hardly of any commercial use in that state. Alloys of iron with carbon are however very widely used. The following table indicates the names given to general groups of such alloys.

Material	Percentage carbon
Wrought iron	0 to 0.05
Steel	0.05 to 2
Cast iron	2 to 4.3

The percentage of carbon alloyed with iron has a profound effect on the properties of the alloy. The term *carbon steel* is used for those steels in which essentially just iron and carbon are present. The term *alloy steel* is used where other elements are included. This chapter considers just the carbon steels. Cast iron is considered in the next chapter.

THE IRON-CARBON SYSTEM

Pure iron at ordinary temperatures has the body-centred cubic structure. This form is generally referred to as α *iron*. The iron will retain this structure up to a temperature of 908°C. At this temperature the structure of the iron changes to face-centred cubic. This is referred to as γ *iron*. At 1388°C this structure changes again to give a body-centred cubic structure known as δ *iron*. The iron retains this form up to the melting point of 1535°C. *Figure 2.1* summarises these various changes.

The name *ferrite* is given to the two body-centred cubic forms of iron, i.e. the α and δ forms. The name *austenite* is given to the face-centred cubic form, i.e. the γ form.

Figure 2.2 shows the iron-carbon system thermal equilibrium diagram. The α iron will accept up to about 0.02% of carbon in solid solution. The γ iron will accept up to 2.0% of carbon in solid solution. With these amounts of carbon in solution the α iron still retains its body-centred cubic structure, and the name ferrite, and the γ iron its face-centred structure, and the name austenite (*Figure 2.3*). The solubilities of carbon in iron, both in the austenite and ferrite forms, varies with temperature. With slow cooling, carbon in excess of that which the α or γ solid solutions can hold at a particular temperature will precipitate. The precipitate is not however as carbon

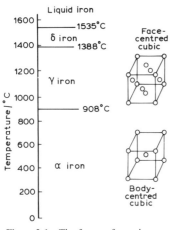

Figure 2.1 The forms of pure iron

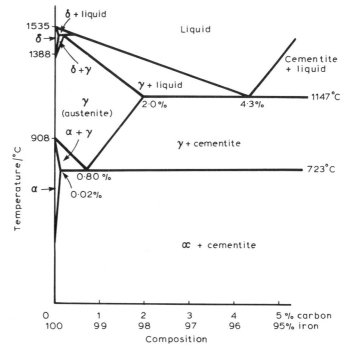

Figure 2.2 The iron-carbon system. (Note: This is really the iron – iron carbide system)

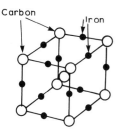

Ferrite, the body centred cubic form, can only accept up to 0·02% carbon in solid solution

(a)

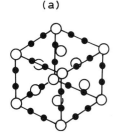

Austenite, the face centred cubic form, can accept up to 2·0% carbon in solid solution

(b)

Figure 2.3 (a) Ferrite, the body centred cubic form, can only accept up to 0.02% carbon in solid solution, (b) Austenite, the face-centred cubic form, can accept up to 2.0% carbon in solid solution

but as iron carbide (Fe_3C), a compound formed between the iron and the carbon. This iron carbide is known as *cementite.* Cementite is hard and brittle.

Consider the cooling from the liquid of an alloy with 0.80% carbon (*Figure 2.4*). For temperatures above 723°C the solid formed is γ iron, i.e. austenite. This is a solid solution of carbon in iron. At

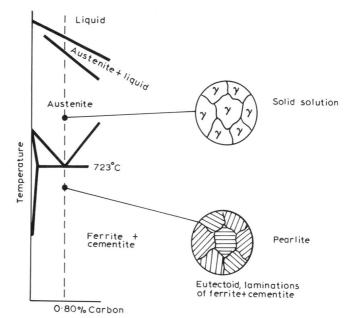

Figure 2.4 Slow cooling of a 0.80% carbon steel

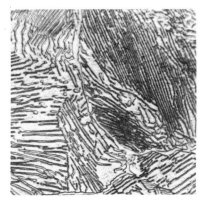

Figure 2.5   Lamellar pearlite × 600 magnification. (From Monks, H. A. and Rochester, D. C., *Technician Structure and Properties of Metals,* Cassell)

723 °C there is a sudden change to give a laminated structure of ferrite plus cementite. This structure is called *pearlite* (*Figure 2.5*). This change at 723 °C is rather like the change that occurs at a eutectic, but there the change is from a liquid to a solid, here the change is from one solid structure to another. This type of change is said to give a *eutectoid*. The eutectoid structure has the composition of 0.8% carbon – 99.2% iron in this case.

Steels containing less than 0.80% carbon are called hypo-eutectoid steels, those with between 0.80% and 2.0% carbon are called hyper-eutectoid steels.

Figure 2.6 shows the cooling of a 0.4% carbon steel, a hypo-eutectoid steel, from the austenite phase to room temperature. When the alloy is cooled below temperature T_1, crystals of ferrite start to grow in the austenite. The ferrite tends to grow at the grain

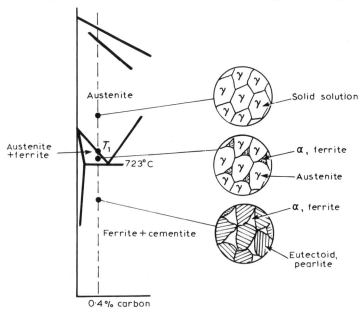

Figure 2.6   Slow cooling of a 0.40% carbon steel

boundaries of the austenite crystals. At 723 °C the remaining austenite changes to the eutectoid structure, i.e. pearlite. The result can be a network of ferrite along the grain boundaries surrounding areas of pearlite (*Figure 2.7*).

Figure 2.8 shows the cooling of a 1.2% carbon steel, a *hyper-eutectoid* steel, from the austenite phase to room temperature. When the alloy is cooled below the temperature T_1, cementite starts to grow in the austenite at the grain boundaries of the austenite crystals. At 723 °C the remaining austenite changes to the eutectoid structure, i.e. pearlite. The result is a network of cementite along the grain boundaries surrounding areas of pearlite.

Thus *hypo*-eutectoid carbon steels consist of a *ferrite* network enclosing pearlite and *hyper*-eutectoid carbon steels consist of a *cementite* network enclosing pearlite.

Example

What would be the expected structure of a 1.6% carbon steel if it was slowly cooled from the austenitic state?

A cementite network enclosing pearlite grains.

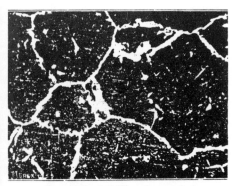

Figure 2.7   A 0.5% carbon steel, slow cooled. Shows network of ferrite.

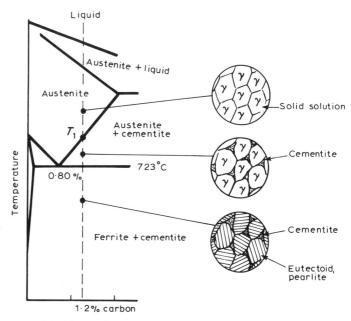

Figure 2.8 Slow cooling of a 1.2% carbon steel

CRITICAL CHANGE POINTS

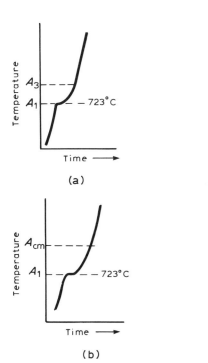

Figure 2.9 Heating curve for a hypo-eutectoid steel, (b) Heating curve for a hyper-eutectoid steel

If water is heated and a graph plotted of temperature with time there will be found to be two discontinuities in the graph where the temperature does not continue to rise at a steady rate despite the heat being supplied at a steady rate. *Figure 1.15* shows one of these discontinuities. They occur at 0°C and 100°C when the state of the water changes. When carbon steels are heated, similar discontinuities occur in the temperature with time graph. The temperatures at which there are changes in the rate of temperature rise, for a constant rate of supply of heat, for steels are known as *arrest points* or *critical points*.

Figure 2.9a shows a heating curve for a hypo-eutectoid steel. The lower critical temperature A_1 is the same for all carbon steels and is the temperature 723°C at which the eutectoid change occurs. For the hypo-eutectoid steel this temperature marks the transformation from a steel with a structure of ferrite and cementite to one of ferrite and austenite. The upper critical temperature A_3 depends on the carbon content of the steel concerned and marks the change from a structure of ferrite and austenite to one solely of austenite. For a 0.4% carbon steel this would be temperature T_1 in *Figure 2.5*.

The heating curve for a hyper-eutectoid steel (*Figure 2.9b*) has the same lower critical temperature A_1 of 723°C, marking the transformation from a steel with a structure of ferrite and cementite to one of austenite and cementite. The upper critical temperature A_{cm} marks the transformation from a structure of austenite and cementite to one solely austenite. For a 1.2% carbon steel this would be temperature T_1 in *Figure 2.8*.

Cooling curves give critical points differing slightly from those produced by heating, cooling giving lower values than heating. The heating critical points are generally denoted by the inclusion of the letter *c* the abbreviation for the French word for heating –*chauffage*, i.e. Ac_1, Ac_3, Ac_{cm}. The cooling critical points are denoted by the

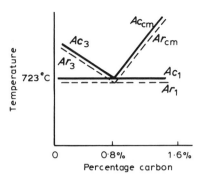

Figure 2.10   Critical points

inclusion of the letter *r* (the abbreviation for the French word for cooling – *refroidissement*), i.e. Ar_1, Ar_3, Ar_{cm}. The letter 'A' used with the critical points stands for the term 'arrest'.

Figure 2.10 shows a graph of the critical point temperatures against the percentage carbon in the steel. The graph is restricted to those carbon percentages that result in steels. The graph is, in fact, the iron-carbon diagram given in *Figure 2.2*. The critical point graph is the thermal equilibrium graph if the heating and cooling graphs were obtained as the result of very slow heating and cooling rates.

THE EFFECT OF CARBON CONTENT ON MECHANICAL PROPERTIES OF STEELS

Ferrite is a comparatively soft and ductile material. Pearlite is a harder and much less ductile material. Thus the relative amounts of these two substances in a carbon steel will have a significant effect on the properties of that steel. *Figure 2.11* shows how the percentages of ferrite and pearlite change with percentage carbon and also how the mechanical properties are related to these changes. The data refers only to steels cooled slowly from the austenitic state.

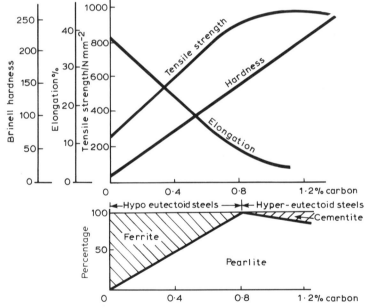

Figure 2.11   The effect of carbon content on the structure and properties of steels

Up to the eutectoid composition carbon steel, i.e. for hypo-eutectoid steels, the decreasing percentage of ferrite and the increasing percentage of pearlite results in an increase in tensile strength and hardness. The ductility decreases, the elongation at fracture being a measure of this. For hyper-eutectoid steels, increasing the amount of carbon decreases the percentage of pearlite and increases the percentage of cementite. This increases the hardness but has little effect on the tensile strength. The ductility also changes little.

Carbon steels are grouped according to their carbon content. *Mild steel* is a group of steels having between 0.10% and 0.25% carbon, *medium-carbon steel* has between 0.20% and 0.50% carbon, *high-carbon steel* has more than 0.5% carbon. Mild steel is a general purpose steel and is used where hardness and tensile strength are not the most important requirements. Typical applications are sections for use as joists in buildings, bodywork for cars and ships, screws, nails, wire. Medium carbon steel is used for agricultural tools, fasteners, dynamo and motor shafts, crankshafts, connecting rods, gears. High carbon steel is used for withstanding wear, where hardness is a more necessary requirement than ductility. It is used for machine tools, saws, hammers, cold chisels, punches, axes, dies, taps, drills, razors. The main use of high carbon steel is thus as a tool steel.

Example

Carbon steel is used for car bumpers. Which type of steel might be suitable?

A car bumper needs to be reasonably ductile so that it will 'give' and so absorb the energy of a collision. It does however have to be reasonably hard so that it is not too easily marked. These arguments would tend to suggest a medium-carbon steel. Another line of argument might be to suggest that car bumpers need to have high strength and hardness so that they can withstand collisions and so prevent damage to the weaker bodywork behind it. This would suggest the need for a high carbon steel.

Car bumpers are generally made from a high carbon steel having about 0.8% carbon, though steel with as little as 0.1% carbon has been used.

Example

Carbon steel is used for the heads of pick-axes. Which type of steel might be suitable?

The pick-axe head will need to be very hard and possess high strength. This would suggest the use of a high-carbon steel.

Pick-axe heads are generally made from a high-carbon steel having about 1% carbon.

RECRYSTALLISATION

When a polycrystalline metal is deformed, perhaps bent or stretched, the grains in the metal become deformed. The metal also becomes harder; this is referred to as *work hardening*. If, however, the metal is heated to a high enough temperature, recrystallisation occurs. New unstrained crystals form from the originally distorted crystals and tend to be uniform in size and comparatively small. The effect of the recrystallisation is a decrease in hardness and strength, and an increase in ductility. If however the metal is kept at the recrystallisation temperature for some time the crystals tend to grow. A coarser grain structure is thus produced. Increasing the size of the grains results in greater ductility at the expense of hardness and strength.

HEAT TREATMENT OF STEEL

Heat treatment can be defined as the controlled heating and cooling of metals in the solid state for the purpose of altering their properties according to requirements. Heat treatment can be applied to steels to alter their properties by changing grain size and the form of the constituents present.

Annealing is the heat treatment used to make a steel softer, and more ductile, remove stresses in the material and reduce the grain size. One form of the annealing process is called *full annealing*. In the case of hypo-eutectoid steels, this involves heating the material to a temperature above the A_3 temperature, holding at that temperature for a period of time and then very slowly cooling it. Typically, the material is heated to about 40°C above the A_3 temperature which has the effect of converting the structure of the steel to austenite. Slow cooling leads to the conversion of the austenite to ferrite and pearlite. The result is a steel in as soft a condition as possible.

A different process has to be used for hyper-eutectoid steels in that heating them to above the A_{cm} temperature turns the entire steel structure into austenite but slow cooling from that temperature results in the formation of a network of cementite surrounding the pearlite. The cementite is brittle and has the effect of making the steel relatively brittle. To make the steel soft the original heating is to a temperature only about 40°C above the A_1 temperature. This converts the structure into austenite plus cementite. Slow cooling from this temperature gives as soft a material as is possible, but not however as soft as the full annealing process gives when applied to hyper-eutectoid steels.

Sub-critical annealing, sometimes referred to as *process annealing,* is often used during cold working processes where the material has to be made more ductile for the process to continue. The process involves heating the material to a temperature just below the A_1 temperature, holding it at that temperature for a period of time and then cooling it at a controlled rate, generally just in air rather than cooling in the furnace as with full annealing. This process leads to no change in structure, no austenite being produced, but just a recrystallisation. The process is used for steels having up to about 0.3% carbon. When sub-critical annealing is applied to steels having higher percentages of carbon the effect of the heating is to cause the cementite to assume spherical shapes, hence the process is often referred to as *spheroidizing annealing*. This spheroidizing results in a greater ductility and an improvement in machinability. *Figure 2.12* summarises the range of annealing processes.

Normalising is a heat treatment process similar to full annealing, and is applied to hypo-eutectoid steels. The steel is heated to about 40°C above the A_3 temperature, held at this temperature for a short while, and then cooled freely in air. The effect of the heating is to form an austenite structure. The cooling rate is however much faster than that with the annealing process. The result is a finer grainer structure, ferrite and pearlite. This finer grain size improves the machinability and gives a slightly harder and greater strength material than that given with full annealing.

Both annealing and normalising involve relatively slow cooling of the steel, but what happens if a steel is cooled very quickly, i.e. quenched? With very slow cooling there is time for diffusion to occur in the solid solutions, with quenching there is not time for such events. If a hypo-eutectoid steel is heated to 40°C above the A_3 temperature all the structure becomes austenitic. Very rapid cooling of this structure does not allow sufficient time for the ferrite structure to be produced, i.e. there is not enough time for the austenite to give up its surplus carbon and so produce ferrite. The result is a new structure called *martensite*. Martensite is a very hard

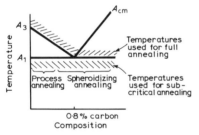

Figure 2.12   Annealing temperatures

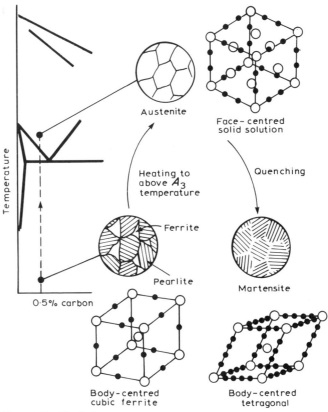

Figure 2.13 Hardening

structure, hence the result of such a process is a much harder
material. The *hardening* process is thus a sequence of heating to
produce an austenitic state and then rapid cooling to produce
martensite (*Figure 2.13*).

If a steel is cooled at, or greater than, a certain minimum rate,
called the *critical cooling rate,* all the austenite is changed into
martensite. This gives the maximum hardness. If the cooling is slower
than this critical cooling rate a less hard structure is produced.

The rate of cooling depends on the quenching medium used.
Water is a commonly used quenching medium and gives a high
cooling rate. However, distortion and cracking may be caused by this
high cooling rate. Oil gives a slower cooling rate, but brine gives a
rate even higher than water. The brine may however give rise to
corrosion problems with the steel. Sodium or potassium hydroxide
solution is sometimes used for very high cooling rates. In order to
minimise the distortion that can be produced during the quenching
process, long items should be quenched vertically, flat sections
edgeways. To prevent bubbles of steam adhering to the steel during
the quenching, and giving rise to different cooling rates for part of
the object, the quenching bath should be agitated. Thick objects
offer special problems in that the outer parts of the object cool more
rapidly than the inner parts. The result can be an outer layer of
martensite and an inner core of pearlite, which gives a variation in
mechanical properties between the inner and outer parts. This effect
is known as the *mass effect.*

Hypo-eutectoid steels are generally hardened by quenching from

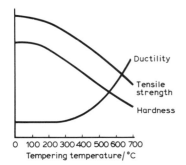

Figure 2.14 The effect of tempering temperature on the properties of a hardened steel

about 40°C above the A_3 temperature, though steels with less than about 0.3% carbon cannot be hardened very effectively. Hyper-eutectoid steels are hardened by quenching from about 40°C above the A_1 temperature. This is because to heat the steel above the A_{cm} temperature and then quench it gives rise to a network of cementite and so a brittle structure.

Tempering is the name given to the process in which a steel, hardened as a result of quenching, is reheated to a temperature below the A_1 temperature in order to modify the structure of the steel. The result is an increase in ductility at the expense of hardness and strength. The degree of change obtained depends on the temperature to which the steel is reheated, the higher the tempering temperature the lower the hardness but the greater the ductility (*Figure 2.14*). By combining hardening with tempering an appropriate balance of mechanical properties can be achieved for a steel.

The following tempering temperatures are used to obtain the required properties for the components concerned.

Tempering temperature/°C	Component
200	Scribers
220	Hacksaw blades
230	Planing and slotting tools
240	Drills, milling cutters
250	Taps, shear blades, dies
260	Punches, reamers
270	Axes, press tools
280	Cold chisels, wood chisels
290	Screw drivers
300	Saws, springs

The lower the temperature at which tempering occurs, the harder the product. Thus scribers need to be hard and have high abrasion resistance. Springs, however, do not need to be so hard but more 'springy'. The scribers may be relatively brittle, the springs will be tougher and much less brittle.

Internal stresses in a material can be relieved to some extent by heating it to a temperature say 50° to 100°C below the A_1 temperature. The material is then usually air cooled from that temperature. *Stress relief* by this method is used with welded components before machining, parts requiring machining to accurate dimensions, castings before machining, etc.

Example

The lower critical point A_1 of a 0.4% carbon steel is 723°C, the upper critical point A_3 being 810°C. To what temperature should the steel be heated for full annealing?

A 0.4% carbon steel is a hypo-eutectoid steel and thus it should be heated to about 40°C above the A_3 temperature, i.e. about 855°C.

Example

The lower critical point A_1 of a 0.5% carbon steel is 723°C, the upper critical point A_3 being 765°C. To what temperature should the steel be heated for normalising?

The required temperature is about 40°C above the A_3 temperature, i.e. about 805°C.

Example

What tempering temperature should be used for a hammer if it is to be hard?

A temperature of about 230°C is generally used. A hard component requires a low tempering temperature.

SURFACE HARDENING OF CARBON STEELS

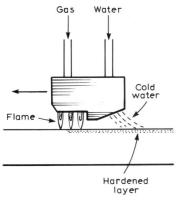

Figure 2.15 A burner with cooling water for flame hardening

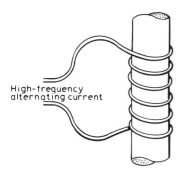

Figure 2.16 Induction hardening of a tube or rod

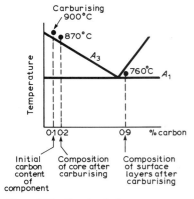

Figure 2.17 Case hardening

There is often the need for the surface of a piece of steel to be hard, i.e. wear resistant, without the entire component being made hard and so often too brittle. Several methods are available for surface hardening. For carbon steels there are two methods:

(1) Selective heating of the surface layers;
(2) Changing the carbon content of the surface layers.

One method of selective heating is called *flame hardening,* which involves heating the surface of a steel with an oxy-acetylene flame and then quenching it before the inner parts of the component have reached the surface temperature (*Figure 2.15*). Another method places the steel component within a coil which carries a high-frequency current, producing currents in the surface layers of the steel by electromagnetic induction (*Figure 2.16*). These currents can heat the surface layers so quickly that the inner parts of the steel component do not reach a high enough temperature for any hardening of them to occur when the component is quenched. This method is called *induction hardening.*

Both these selective heating methods require the surface layers of the steel to be brought to above the A_3 temperature while the inner parts of the component remain at a temperature significantly lower. The A_3 temperature results in the surface layers becoming austenitic. When the material is quenched the surface layers become martensite, a much harder material than the inner parts of the component for which the change has not occurred. Steels for surface hardening by these methods require more than about 0.4% carbon content. The hardness produced is typically about 600 HB.

Low-carbon steels can be surface hardened by increasing the carbon content of the surface layers. This method is generally known as *case hardening.* There are a number of stages in this process. In the first stage, *carburising,* the carbon is introduced into the surface layers. With *pack carburising* the steel component is heated to above the A_3 temperature, say 900°C for a 0.1% carbon alloy, while packed in charcoal and barium carbonate. The length of time the component is kept at this temperature determines the depth to which the carbon penetrates the surface, this being referred to as the depth of case. In *gas carburising* the component is heated to above the A_3 temperature in a furnace in an atmosphere of a carbon-rich gas. With *cyanide carburising* the component is heated in a bath of liquid sodium cyanide. The results by all these methods is an increase in carbon in the surface layers of the steel component. Thus while the surface layers may have, say, 0.9% carbon the inner core might be 0.2%, some carbon often reaching the core.

The lengthy heating during the carburising treatment causes grain growth, both in the core and the surface layers. Because of the different carbon contents of the core and the surface layers, two heat treatment processes are needed to refine the grains in the core and the surface layers. The core is first refined. The component is heated to above the A_3 temperature for the carbon content of the core (*Figure 2.17*). For a core with 0.2% carbon this is a temperature of about 870°C. The component is then quenched in oil. This treatment results in a fine grain core. The surface layers after this treatment are, however, rather coarse, brittle martensite, because the 870°C temperature is well above the critical temperature for a 0.9% carbon steel. A second heat treatment is then given, specifically to refine the surface layers. The component is heated to above the A_1

temperature, i.e. about 760°C, and then water quenched. The result is fine martensite, which is hard in the surface, while the core is tempered by the treatment. A low temperature tempering is then usually employed about 150°C, to relieve internal stresses produced by the treatment.

The case hardening treatment might thus be:

(1) Carburise by heating to 950°C in a carbon-rich environment,
(2) Core refine by heating to 870°C and quench in oil,
(3) Refine the surface layers, and so case harden, by heating to 750°C and quench in water,
(4) Temper at 150°C to relieve internal stresses.

The hardness produced by such a method is typically of the order of 850 HB.

HEAT TREATMENT EQUIPMENT

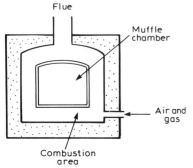

Figure 2.18 The basic form of a gas-fired muffle furnace

The industrial heat treatment of metals requires:

(1) Furnaces to heat the components to the required measured temperature;
(2) Quenching equipment.

An important consideration before the choice of furnace is made is whether the atmosphere in which the components are heated is to be oxidising, i.e. generally air, or some inert gas to prevent oxidisation. A *muffle furnace* (*Figure 2.18*) is one in which the components are contained in a chamber, the so-called muffle chamber, and the heat supplied externally to the chamber. This enables the atmosphere in the chamber to be controlled. A non-muffle furnace has no control over the atmosphere surrounding the components.

Heating an iron alloy in an oxidising atmosphere can result in the carbon in the outer layers of the material combining with the oxygen to give carbon dioxide, which escapes with the flue gases and so leaves the outer layers of the alloy with less carbon than the inner layers. This effect is known as *decarburisation* (see page 41 for an application of this effect). Scaling and tarnishing can also occur when an alloy is heated in an oxidising atmosphere. Heating in a non-oxidising atmosphere can eliminate costly cleaning processes.

The *salt-bath furnace* consists essentially of a pot containing a molten salt, which may be heated directly by an electric current being passed through it or by external heating of the pot. Salt-bath furnaces give an even and rapid heating of components immersed in the salt. By suitable choice of the salt, carburisation can be promoted and so surface hardening occur. No oxidisation takes place for the components immersed in the salt as air is excluded.

Quenching may involve the immersion of the hot components in water, brine, oil or sodium or potassium hydroxide solution. The quenching tanks are designed so that the quenching liquid remains, as nearly as possible, at a constant temperature. In the case of brine or oil baths, these may stand in large tanks through which water flows, the water carrying the heat away and so keeping the quenching liquids at a near constant temperature during the quenching.

There are a number of hazards during heat treatment processes:

(1) Hot metals are being handled, so care should be taken in approaching any piece of metal during the heat treatment process.
(2) Quenching baths and gas fired furnaces give off dangerous

fumes. The heat treatment area needs to be well ventilated and fume extraction systems should be used.

(3) Fire hazards are present with furnaces and also oil quenching tanks. The immersion of a piece of hot metal in oil leads to some of the oil being vaporised and the possibility of a fire. If this happens the air supply to the tank should be cut off by covering the tank with a lid.

(4) Explosions are possible with gas-fired furnaces if the furnace becomes charged with a gas-air mixture before the lighter is operated. The lighter should be operated before the gas is introduced into the combustion chamer.

(5) Some of the materials used in heat treatment are poisonous, e.g. sodium cyanide is used for salt-bath furnaces.

PROPERTIES OF CARBON STEELS

The table below shows the properties that might be obtained with normalised carbon steels. The greater the tensile strength the greater the forces a given cross-section of the material can withstand before breaking. The *tensile strength* is defined as the maximum force a material can withstand divided by the initial cross-sectional area of the material. The *percentage elongation* is defined as the percentage change in length of a sample after breaking, i.e. (final gauge length minus initial gauge length) $\times$ 100/initial gauge length. The greater the elongation the more ductile the material. The *hardness* of a material can be defined in terms of the ability of the material to withstand indentation. One form of the test is the Brinell hardness test. This presents hardness by numbers, denoted by HB. The larger the HB number the harder the material.

Composition	Condition	Tensile strength /N mm^{-2}	Elongation /%	Hardness /HB
0.2% carbon	Normalised	450	34	130
0.4% carbon	Normalised	580	25	165
0.6% carbon	Normalised	720	18	220
0.8% carbon	Normalised	920	12	260

A normalised carbon steel has a tensile strength approximately 20% higher than the same steel when annealed. The annealed material does however have greater ductility but lower hardness. Steels with less than about 0.3% carbon cannot be hardened effectively. Hardening and tempering a 0.50% carbon steel will raise the tensile strength by about 40 to 50% and the hardness by about 10%, the elongation being reduced slightly.

In addition to carbon, all the above alloys contain manganese and other elements. The effect of the manganese is to improve the microstructure of the alloy and also make it easier to harden.

The following extract from the data sheets of a manufacturer illustrates the form in which mild steel strip may be obtained (Courtesy of Arthur Lee & Sons Ltd, Sheffield).

Cold-rolled steel strip

Quality	New Classification	C %	Mn %	S %	P %
Extra deep drawing (non-ageing)	CS1	< 0.07	< 0.45	< 0.030	< 0.025
Extra deep drawing	CS2	< 0.08	< 0.45	< 0.035	< 0.030
Deep drawing	CS3	< 0.10	< 0.50	< 0.040	< 0.040
Ordinary quality	CS4	< 0.12	< 0.50	< 0.050	< 0.050

Size range
(1) Coils: 455 mm wide and narrower × 3.70 mm and thinner, down to 0.10 mm.
(2) Lengths: as for coils but with minimum gauge of 0.45 mm.
Maximum length for strip in cut lengths is normally 3.6 m for all sizes.

Tempers
CS4 Hard
For maximum strength where no bending or drawing is involved. For clean and crisp shearing or blanking. Tensile strength >540 N/mm².
CS4 Half Hard
Strip will stand a 90° transverse bend over a radius equal to its own thickness. Useful for simple bending operations. Tensile strength 420 to 540 N/mm².
CS4 Quarter Hard
Strip will stand a close bend in the transverse direction and a 90° longitudinal bend over a radius equal to its own thickness. Tensile strength 350 to 420 N/mm².
CS4 Annealed (Dead Soft)
Strip will stand being bent flat on itself in both directions, and can also be used for simple pressing and drawing. Tensile strength <350 N/mm².
CS3 Deep drawing quality
Suitable for drawing and pressing, where ductility is an essential requirement. Tensile strength not specified.
CS2 Extra deep drawing quality
Intended for the most arduous and complicated drawing operations. Tensile strength not specified.

PROBLEMS
1. Sketch and label the steel section of the iron-carbon system, using the terms austenite, ferrite and cementite.

2. How do the structures of *hypo*-eutectoid and *hyper*-eutectoid steels differ at room temperature as a result of them being slowly cooled from the austenitic state?

3. Describe the form of the microstructure of a slowly cooled steel having the eutectoid structure.

4. What would be the expected structure of a 1.1% carbon steel if it were cooled slowly from the austenitic state?

5. Explain the terms critical points A_1, A_3 and A_{cm}.

6. Explain how the percentage of carbon present in a carbon steel affects the mechanical properties of the steel.

7. Carbon steel is used for the following items. Which type of carbon steel would be most appropriate for each item?
(a) Railway track rails.
(b) Ball bearings.
(c) Hammers.
(d) Reinforcement bars for concrete work.
(e) Knives.

8. Describe the following heat treatments applied to steels.
(a) Full annealing.

(b) Normalising.
(c) Hardening.
(d) Tempering.

9. What will be the form of the microstructure of a 0.5% carbon steel after the following treatment: Heat and soak at 805°C and then a very slow cool to room temperature?

10. How would the answer to Problem 9 have differed if the steel, instead of being slowly cooled from 805°C to room temperature, had been quenched?

11. Explain how a 0.4% carbon steel would be hardened. Give details of the temperatures involved.

12. Describe the difference between full annealing and normalising.

13. State what is meant by the critical cooling rate.

14. How does increasing the temperature at which a carbon steel is tempered change the final properties of the steel?

15. How would the mechanical properties of 0.6% carbon steels differ if the following heat treatments were applied?
    (a) Heat and soak at 800°C and then quench in cold water.
    (b) Heat and soak at 800°C and then quench in oil.
    (c) Heat and soak at 800°C and then slowly cool in the furnace.

16. A carbon steel with 1.1% carbon is to be given a full annealing treatment. What temperature and cooling rate are necessary for such a treatment?

17. Why are cylindrical objects quenched vertically?

18. A cold chisel is tempered at a temperature of 280°C while a scriber is tempered at 200°C. How does the hardness of the steel differ for the two items as a result of the differing tempering temperatures? Why are the components required to have different hardnesses? What would happen if there was an error and the tempering temperature for the cold chisel was as high as 380°C?

19. Explain, by reference to an iron-carbon thermal equilibrium diagram, the procedures used for the case hardening of a low carbon steel.

20. Describe the functions of the muffle and salt-bath furnaces?

21. State the form of heat treatments needed to effect the following changes:
    (a) a 0.2% carbon steel to be made as soft as possible;
    (b) a 1.0% carbon steel to be made as soft as possible;
    (c) a 0.4% carbon steel to be made as hard as possible;
    (d) a 0.2% carbon steel to be case hardened.

3 Cast irons

Objectives: At the end of this chapter you should be able to:
Describe the structure and properties of cast irons.
Select cast irons for specific applications.

COMPOSITION

The term cast iron arises from the method by which cast iron is produced. Pig iron is remelted in a furnace and the properties of the iron modified by additions of other materials such as steel scrap. The resulting iron alloy is cast, hence the term cast iron. Cast irons have more than about 2% carbon and often significant amounts of silicon as well as smaller amounts of other materials. Cast irons are used widely because of their ease of melting and hence use where components are to be produced by casting. Cast irons are also relatively cheap.

The structure, and hence physical properties, of cast irons is affected by:

(a) the carbon content,
(b) the rate of cooling of the iron from the liquid state to room temperature,
(c) the presence of other elements.

In carbon steels the carbon exists in the form of cementite, a compound formed between iron and carbon. In cast iron the carbon occurs as graphite or cementite. Graphite is a soft, grey substance used as the 'lead' in pencils. Cementite, however, is very hard. The higher the graphite content of an iron the more grey the appearance of the iron. The various cast irons can be classified by the presence or otherwise of the carbon as graphite and the type of structure of the iron. *Figure 3.1* shows the general classification. The formation of

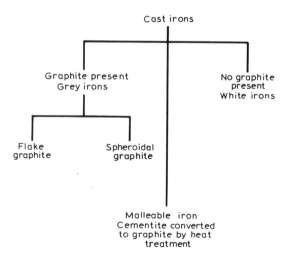

Figure 3.1 A general classification of cast irons

graphite in a cast iron is affected by the rate of cooling in that a high rate of cooling tends to hinder the formation of graphite and so give a structure with more cementite. The presence of other elements can also have a significant effect on the production of graphite.

THE CASTING PROCESS

The casting process involves the heating of a metal to make it liquid, pouring it into a mould, and then permitting it to solidify. The result is a piece of metal shaped to the configuration of the mould. The requirements for the metal used for a casting are that it should have a relatively low melting point so that it can be liquefied easily; it should then have fluidity, so that it can be poured easily into the mould and flow quickly to all parts of it; it should not shrink too much when it solidifies and cools; the cold metal should have reasonable strength. The metals most used for casting are aluminium, brass, bronze, magnesium, steel and iron. Iron is used a considerable amount for casting because of its relatively low melting point, fluidity when liquid, low shrinkage and strength when cold.

GREY AND WHITE CAST IRONS

Figure 3.2 shows that part of the iron-carbon diagram of relevance to cast irons, i.e. iron alloys containing more than 2.0% carbon. At a temperature of about 1147°C a eutectic occurs for alloys with 4.3% carbon.

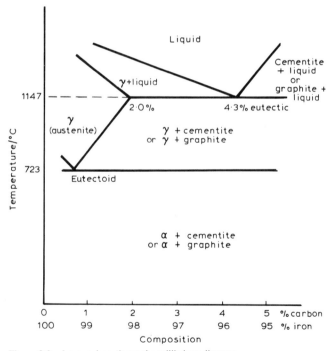

Figure 3.2 Iron-carbon thermal equilibrium diagram

If we consider the cooling of an iron from the liquid state we have the following sequence of events. With very slow cooling, when solidification starts to occur the result is the formation of austenite in the liquid. At 1147°C the liquid solidifies to give austenite plus graphite. As the temperature is further lowered the graphite grows due to precipitation from the austenite. At 723°C the remaining austenite changes to ferrite plus graphite. The result at room temperature is a structure of ferrite and graphite. This type of cast iron is known as a *grey iron* because of the grey appearance of its freshly fractured surface. This grey iron is referred to as a ferritic form of grey iron.

A faster cooling rate can lead to a structure in which the austenite changes to a mixture of ferrite and pearlite at 723°C, or with an even faster cooling rate, entirely pearlite. The results are structures involving graphite with either ferrite and pearlite or just pearlite. The greater the amount of pearlite in these grey irons the harder the material.

When the iron solidifies from the liquid state the graphite usually forms as flakes. Hence the structure at room temperature has graphite flakes in ferrite, ferrite and pearlite, or pearlite, depending on the rate of cooling (*Figure 3.3*).

Figure 3.3   A grey cast iron. The structure consists of black graphite flakes in a matrix of pearlite. (Courtesy of BCIRA)

With even faster cooling a different type of structure is formed. The solidification at 1147°C gives austenite and cementite. As the temperature is further lowered the cementite grows due to precipitation from the austenite. At 723°C the remaining austenite changes to pearlite. The result at room temperature is a structure of cementite and pearlite (*Figure 3.4*). This type of cast iron is known as a *white iron* because of the white appearance of its freshly fractured surface. White iron, because of its high cementite content, is hard and brittle. This makes it difficult to machine and hence of limited use. The main use is where a wear-resistant surface is required.

A casting will often have sections with differing thicknesses. The rate of cooling of a part of a casting will depend on the thickness of the section concerned, the thinner the section the more rapidly it will cool. This means that there is likely to be a variation in the properties throughout the casting, the thinner section possibly giving white iron

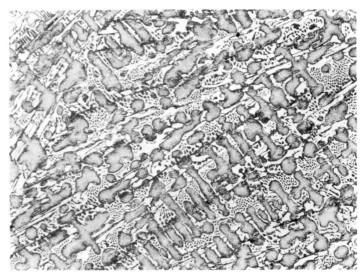

Figure 3.4   A white cast iron. The structure consists of white cementite and dark pearlite. (Courtesy of BCIRA)

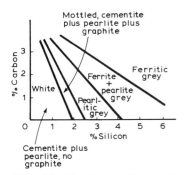

Figure 3.5   The effect of silicon on the structure of cast iron during normal cooling

while the thicker part gives a grey iron. There is thus a need to consider carefully the way in which a casting cools if the properties of the resultant casting are to be controlled.

Some elements when included with the carbon in cast iron promote the formation of graphite, others promote the production of the carbide. Silicon and nickel promote the formation of graphite by causing the cementite to become unstable and precipitate carbon. Thus the effect of including silicon with the carbon is to increase the chance of a grey iron being produced, rather than a white iron. *Figure 3.5* illustrates this. With, say, a 3% carbon iron and normal cooling we can have the following situations:

(a) With no silicon a white iron is produced.
(b) With 1% silicon a pearlitic grey iron is produced.
(c) With 2% silicon a grey iron with both ferrite and pearlite is produced.
(d) With 3% silicon a ferritic grey iron is produced.

The type of iron produced depends on the rate of cooling as well as the silicon content. The silicon content can however be chosen for a particular casting, which will cool at some particular rate, so that the required type of cast iron is produced.

Sulphur often exists in cast iron and has the effect of stabilising cementite, i.e. the opposite effect to that of silicon. The presence of sulphur thus favours the production of white iron rather than grey iron. The sulphur thus increases the hardness of the cast iron and also increases the brittleness.

The addition of small amounts of manganese to a sulphur containing iron enables the sulphur to combine with the manganese to form manganese sulphide. This removal of the free sulphur has the effect of increasing the chance of grey iron being produced. With higher amounts of manganese a reaction occurs between the manganese and the carbon with the production of manganese carbide which considerably hardens the cast iron.

The addition of phosphorus to the iron has little effect on the amounts of cementite or graphite but does increase the fluidity of the iron in casting. It does this by reacting with the iron to produce iron phosphide which has a low melting point.

The following is a typical type of composition for a grey iron:

Carbon 3.2 to 3.5% (most as graphite)
Silicon 1.3 to 2.3%
Sulphur 0.10%
Managanese 0.5 to 0.7%
Phosphorus 0.15 to 1.0%
A typical white iron could have the following composition:
Carbon 3.3% (all in cementite)
Silicon 0.5%
Sulphur 0.15%
Manganese 0.5%
Phosphorus 0.5%

Example

How does white cast iron differ from grey cast iron in (a) microstructure and (b) hardness?

(a) White cast iron has, at room temperature, all the carbon combined with the iron as cementite. Grey cast iron, however, has most of its carbon as graphite.

(b) The effect of the cementite is to give a hard iron, virtually unmachineable. The presence of the graphite results in a soft iron. Thus white cast irons are hard while grey cast irons are relatively soft.

Example

Part of a casting is deliberately cooled much more rapidly than other parts. What are the effects of such a treatment?

The higher the rate of cooling the more likely is the formation of white iron, slow rates of cooling lead to grey iron. The deliberate fast cooling of part of a casting can be for the purpose of producing a very hard part. This may be because a wear resistant part is required.

HEAT TREATMENT OF GREY IRONS

Annealing is used with grey cast irons to provide optimum machineability and remove stresses. Annealing involves heating the cast iron to about 40°C above the A_1 temperature, i.e. about 760°C, soaking at that temperature and then very slow cooling.

Grey cast iron can be hardened by quenching. Such a treatment is generally followed by tempering. A tempering temperature of 475°C is common.

A stress relieving treatment is often used before a significant amount of machining takes place. Such a treatment involves heating to about 550°C, i.e. below the A_1 temperature.

MALLEABLE CAST IRONS

Malleable cast irons are produced by the heat treatment of white cast irons. Three forms of malleable iron occur: *whiteheart, blackheart* and *pearlitic*. Malleable irons have better ductility than grey cast irons and this combined with their high tensile strengths makes them a useful material.

In the blackheart process, white iron castings are heated in a non-oxidising atmosphere to 900°C and soaked at that temperature for two days or more. This causes the cementite to break down. The

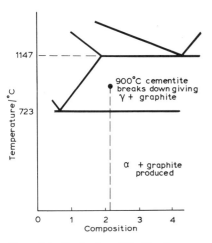

Figure 3.6 Production of blackheart malleable iron

result is spherical aggregates of graphite in austenite. The casting is then cooled very slowly, resulting in the austenite changing into ferrite and more graphite (*Figure 3.6*). *Figure 3.7* shows the form of the product.

The whiteheart process also involves heating white iron castings to about 900°C and soaking at that temperature. But in this process the castings are packed in canisters with haematite iron ore. Thi gives an oxidising atmosphere. Where the casting is thin, the carbon is oxidised forming a gas and so leaves the casting. In the thicker sections of the casting only the carbon in the surface layers leaves. The result, after very slow cooling, is a ferritic structure in the thin sections of the casting and, in the thick sections, a ferrite outer layer with a ferrite plus pearlite inner core.

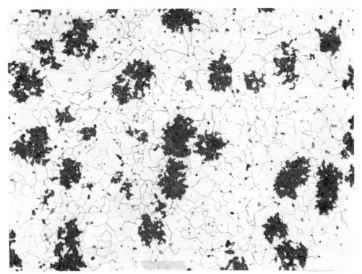

Figure 3.7 Blackheart malleable iron. The structure consists of 'rosettes' of graphite in a matrix of ferrite. (Courtesy of BCIRA)

Pearlitic malleable iron is produced by heating a white iron casting in a non-oxidising atmosphere to 900°C and then soaking at that temperature. This causes the cementite to break down and give spherical aggregates of graphite in austenite, as with blackheart iron. However, if a more rapid cooling is used a pearlitic structure is produced. This pearlitic malleable iron has a higher tensile strength than blackheart iron.

Pearlitic malleable irons can be produced by other methods. One method involves adding 1% manganese to the iron. The manganese inhibits the production of graphite. The higher strength pearlitic malleable irons are produced by quenching the iron from 900°C and then tempering.

NODULAR CAST IRONS

Nodular iron, or *spheroidal-graphite (SG) iron* or *ductile iron* as it is sometimes called, has the graphite in the iron in the form of nodules of spheres. Magnesium or cerium is added to the iron before casting occurs. The effect of these materials is to prevent the formation of graphite flakes during the slow cooling of the iron, the graphite forms nodules instead. At room temperature the structure of the cast

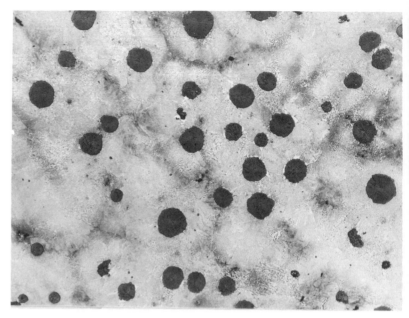

Figure 3.8   Spheroidal graphite iron. The structure consists of black spheroidal graphite in a matrix of pearlite and white cementite. (Courtesy of BCIRA)

iron is mainly pearlitic with nodules of graphite (*Figure 3.8*). The resulting material is more ductile than a grey iron.

A heat treatment process can be applied to a pearlitic nodular iron to give a microstructure of graphite nodules in ferrite. The treatment is to heat to 900°C, soak at that temperature and then slowly cool. This ferritic form is more ductile, but has less tensile strength, than the pearlitic form.

PROPERTIES AND USES OF CAST IRONS

The table shows the types of properties that might be obtained with cast irons. Page 33 explains the terms used in the table.

Type of material	Condition	Tensile strength /N mm^{-2}	Elongation / %	Hardness / HB
Grey iron	As cast	150–400	0.5–0.7	130–300
White iron	As cast	230–460	0	400
Blackheart	Annealed	290–340	6–12	125–140
Whiteheart	Annealed	270–410	3–10	120–180
Pearlitic malleable	Normalised	440–570	3– 7	140–240
Nodular ferritic	As cast	370–500	7–17	115–215
Nodular pearlitic	As cast	600–800	2– 3	215–305

With the exception of the white iron, all the other cast irons listed above give good to very good machining. Similarly for casting, the white iron gives only fair castings while the other cast irons give good to very good castings. Fair welds can be achieved with the cast irons, apart from the white iron for which welding is poor. White iron has a very high abrasion resistance.

The following are typical uses of the various cast irons:

Grey iron	Water pipes, motor cylinders and pistons, machine castings, crankcases, machine tool beds, manhole covers
White iron	Wear resistant parts, such as grinding mill parts, crusher equipment
Blackheart	Wheel hubs, pedals, levers, general hardware, brake shoes
Whiteheart	Wheel hubs, bicycle and motor-cycle frame fittings
Pearlitic malleable	Camshafts, gears, couplings, axle housings
Nodular ferritic	Heavy duty piping
Nodular pearlitic	Crankshafts

In selecting a cast iron for a particular application the following considerations are taken into account:

Grey irons	Very good machineability and stability. Able to damp out vibrations, i.e. considered to have a good damping capacity. Excellent wear resistance due to the graphite giving a self-lubricating effect for metal-to-metal contacts. Relatively poor tensile strength and ductility, but good strength in compression. Not good for shock loads.
White iron	Excellent abrasion resistance. Very hard. Virtually unmachineable so has to be cast to the required shape and dimensions
Malleable iron	Good machineability and stability. Higher tensile strength and ductility than grey iron. Better for shock loads than grey iron
Nodular iron	High tensile strengths with reasonable ductility. Good machineability and wear characteristics, but not as good as grey iron. Better for shock loads than grey iron (about the same as malleable iron)

The following extracts from a manufacturer's information sheets illustrate the types of product that can be produced with cast irons. (Courtesy of Stirling Metals Ltd).

Cylinder Blocks and Transmission Cases in High Duty Grey Iron

The technology involved in the design and production of modern power units requires materials and components to be of the very highest grades. Thus the cylinder block, frequently of complicated and intricate design, must be completely sound in structure to withstand the mechanical stresses and pressures which are generated during service.

For over sixty years, the iron foundries of Sterling Metals have specialised in the production of cylinder blocks in grey iron for internal combustion engines.

Considerable attention has necessarily been devoted to the development of the correct material, which must have hard-wearing properties, evenness of grain structure through thick and thin sections, freedom from porosity and brittleness, and be capable of machining at high speeds. The following are details and properties of the normal specification:

Sterling Metals Cylinder Iron
BSS 1452 Grades 14 and 17
Chemical properties

Total carbon	3.20–3.40%
Silicon	2.10–2.30%
Manganese	0.60–0.90%
Chromium	0.20–0.40%
Combined carbon	0.55–0.75%
Phosphorus	0.15% max.
Sulphur	0.12% max.

Mechanical properties
Brinell Hardness 180–240
Tensile strength 230–280 N/mm^2. Nominal diameter of test bar as cast 30.48 mm.

Ley's malleable castings solve tough problems

A leading manufacturer wanted a planet carrier for a big tractor shovel. A planet carrier which was fairly complex in design, had exceptional shock resistance and durability and which could be easily machined.

They came to the right people — Ley's, whose Black Heart Ferritic Malleable Iron has all those properties.
(Courtesy of Ley's Malleable Castings Company Limited).

PROBLEMS

1. Explain the effects on the microstructure and properties of cast iron of (a) cooling rate, (b) carbon content, (c) the addition of silicon, manganese, sulphur and phosphorus.

2. Describe the conditions under which (a) grey iron, (b) white iron are produced.

3. In what way does the section thickness of a casting affect the structure of the cast iron?

4. How do the mechanical properties of malleable irons compare with those of grey irons?

5. Describe the way in which the structures of nodular irons and malleable irons differ from those of grey irons?

6. Which types of cast iron have high ductilities?

7. Which types of cast irons have high tensile strengths?

8. Describe the forms of the microstructure of (a) blackheart and (b) whiteheart cast irons.

9. Why is it important to know whether a casting will require any machining before deciding on the material to be used?

10. How does the presence of graphite as flakes or nodules affect the properties of the cast iron?

11. Which type of cast iron would be most suitable for a situation where there was a high amount of wear anticipated?

12. Which types of cast iron would you suggest for the following applications? Justify your answers.
   (a) Sewage pipe.
   (b) Crankshaft in an internal combustion engine.
   (c) Brake discs.
   (d) Manhole cover.

4 Non-ferrous alloys

Objectives: At the end of this chapter you should be able to:
Describe the general properties of (a) aluminium and its alloys, (b) copper and its alloys, (c) magnesium and its alloys, (d) nickel and its alloys, (e) titanium and its alloys, (f) zinc and its alloys.
Describe typical applications of the above range of elements and alloys.
Select materials for specific applications.

THE RANGE OF ALLOYS

The term *ferrous alloys* is used for those alloys having iron as the base element, e.g. cast iron and steel. The term *non-ferrous alloys* is used for those alloys which do not have iron as the base element, e.g. alloys of aluminium. The following are some of the non-ferrous alloys in common use in engineering:

Aluminium alloys	Aluminium alloys have a low density, good electrical and thermal conductivity, high corrosion resistance. Typical uses are metal boxes, cooking utensils, aircraft body-work and parts.
Copper alloys	Copper alloys have good electrical and thermal conductivity, high corrosion resistance. Typical uses are pump and valve parts, coins, instrument parts, springs, screws. The names brass and bronze are given to some forms of copper alloys.
Magnesium alloys	Magnesium alloys have a low density, good electrical and thermal conductivity. Typical uses are castings and forgings in the aircraft industry.
Nickel alloys	Nickel alloys have good electrical and thermal conductivity, high corrosion resistance, can be used at high temperatures. Typical uses are pipes and containers in the chemical industry where high resistance to corrosive atmospheres is required, food processing equipment, gas turbine parts. The names Monel, Inconel and Nimonic are given to some forms of nickel alloys.
Titanium alloys	Titanium alloys have a low density, high strength, high corrosion resistance, can be used at high temperatures. Typical uses are in aircraft for compressor discs, blades and casings, in chemical plant where high resistance to corrosive atmospheres is required
Zinc alloys	Zinc alloys have good electrical and thermal conductivity, high corrosion resistance, low melting points. Typical uses are as car door handles, toys, car carburettor bodies—components that in general are produced by die casting

Non-ferrous alloys have, in general, these advantages over ferrous alloys:

(a) good resistance to corrosion without special processes having to be carried out;

(b) most non-ferrous alloys have a much lower density and hence lighter weight components can be produced;

(c) casting is often easier because of the lower melting points;

(d) cold working processes are often easier because of the greater ductility;

(e) high thermal and electrical conductivities;

(f) more decorative colours.

Ferrous alloys have these advantages over non-ferrous alloys:

(a) generally greater strengths;

(b) generally stiffer materials, i.e. larger values of Young's modulus;

(c) better for welding.

ALUMINIUM

Pure aluminium has a density of $2.7 \, 10^3 \, kg \, m^{-3}$, compared with that of $7.9 \, 10^3 \, kg \, m^{-3}$ for iron. Thus for the same size component the aluminium version will be about one third of the mass of an iron version. Pure aluminium is a weak, very ductile, material. It has an electrical conductivity about two thirds that of copper but weight for weight is a better conductor. It has a high thermal conductivity. Aluminium has a great affinity for oxygen and any fresh metal in air rapidly oxidises to give a thin layer of the oxide on the metal surface. This surface layer is not penetrated by oxygen and so protects the metal from further attack. The good corrosion resistance of aluminium is due to this thin oxide layer on its surface.

High purity aluminium is too weak a material to be used in any other capacity than a lining for vessels. It is used in this way to give a high corrosion resistant surface. High purity aluminium is 99.5%, or more, aluminium.

Commercial purity aluminium, 99.0 to 99.5% aluminium, is widely used as aluminium foil for sealing milk bottles, thermal insulation, as kitchen foil for cooking. The presence of a relatively small percentage of impurities in aluminium considerably increases the tensile strength and hardness of the material.

The mechanical properties of aluminium depend not only on the purity of the aluminium but also upon the amount of work to which it has been subject. The effect of working the material is to fragment the grains. This results in an increase in tensile strength and hardness and a decrease in ductility. By controlling the amount of working different degrees of strength and hardness can be produced. These are said to be different *tempers*. The properties of aluminium may thus, for example, be referred to as that for the annealed condition, the half-hard temper and the fully hardened temper.

PROPERTIES OF ALUMINIUM

The table shows the types of properties that might be obtained with aluminium. Page 33 explains the terms used in the table.

Composition %	Condition	Tensile strength / N mm^{-2}	Elongation %	Hardness / HB
99.99	Annealed	45	60	15
	Half hard	82	24	22
	Full hard	105	12	30
99.8	Annealed	66	50	19
	Half hard	99	17	31
	Full hard	134	11	38
99.5	Annealed	78	47	21
	Half hard	110	13	33
	Full hard	140	10	40
99	Annealed	87	43	22
	Half hard	120	12	35
	Full hard	150	10	42

ALUMINIUM ALLOYS

Aluminium alloys can be divided into two groups:

(1) Wrought alloys,
(2) Cast alloys.

Each of these groups can be divided into two further groups:

(a) Those alloys which are not heat treatable,
(b) Those alloys which are heat treated.

The term *wrought material* is used for a material that is suitable for shaping by working processes, e.g. forging, extrusion, rolling. The term *cast material* is used for a material that is suitable for shaping by a casting process.

The non-heat treatable wrought alloys of aluminium do not significantly respond to heat treatment but have their properties controlled by the extent of the working to which they are subject. A range of tempers is thus produced. Common alloys in this category are aluminium with manganese or magnesium. A common aluminium-manganese alloy has 1.25% manganese, the effect of this manganese being to increase the tensile strength of the aluminium. The alloy still has a high ductility and good corrosion properties. This leads to uses such as kitchen utensils, tubing and corrugated sheet for building. Aluminium-magnesium alloys have up to 7% magnesium. The greater the percentage of magnesium the greater the tensile strength (*Figure 4.1*). The alloy still has good ductility. It has excellent corrosion resistance and thus finds considerable use in marine environments, e.g. constructional materials for boats and ships.

The heat treatable wrought alloys can have their properties changed by heat treatment. Copper, magnesium, zinc and silicon are common additions to aluminium to give such alloys. *Figure 4.2*

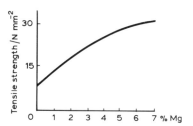

Figure 4.1 The effect on the tensile strength of magnesium content in annealed aluminium-magnesium alloys

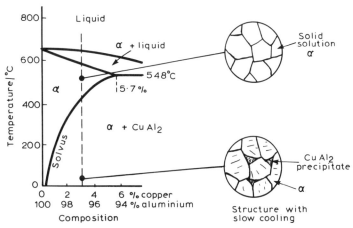

Figure 4.2 Thermal equilibrium diagram for aluminium-copper alloys

shows the thermal equilibrium diagram for aluminium-copper alloys. When such an alloy, say 3% copper – 97% aluminium, is slowly cooled the structure at about 540°C is a solid solution of the α phase. When the temperature falls below the solvus temperature a copper-aluminium compound is precipitated. The result at room temperature is α solid solution with this copper-aluminium compound precipitate ($CuAl_2$). The precipitate is rather coarse, but this structure of the alloy can be changed by heating to about 500°C, soaking at that temperature, and then quenching, to give a supersaturated solid solution, just α phase with no precipitate. This treatment, known as *solution treatment,* results in an unstable situation. With time a fine precipitate will be produced. Heating to, say, 165°C for about ten hours hastens the production of this fine precipitate (*Figure 4.3*). The microstructure with this fine precipitate is both stronger and harder. The treatment is referred to as *precipitation hardening* (see page 19 for more information).

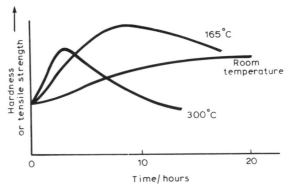

Figure 4.3   The effects of time and temperature on hardness and strength for an aluminium-copper alloy

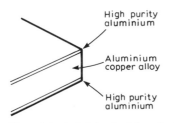

Figure 4.4   A clad duraluminium sheet

A common group of heat treatable wrought alloys is based on aluminium with copper. Thus one form has 4.0% copper, 0.8% magnesium, 0.5% silicon and 0.7% manganese. This alloy is known as Duralumin. The heat treatment process used is solution treatment at 480°C, quenching and then precipitation hardening at either room temperature for about 4 days or 10 hours at 165°C. This alloy is widely used in aircraft bodywork. The presence of the copper does however reduce the corrosion resistance and thus the alloy is often clad with a thin layer of high purity aluminium to improve the corrosion resistance (*Figure 4.4*).

The precipitation hardening of an aluminium-copper alloy is due to the precipitate of the aluminium-copper compound. The age hardening of aluminium-copper-magnesium-silicon alloys is due to the precipitates of both an aluminium-copper compound $CuAl_2$ and an aluminium-copper-magnesium compound $CuAl_2Mg$. Other heat treatable wrought alloys are based on aluminium with magnesium and silicon. The age hardening with this alloy is due to the precipitate of a magnesium-silicon compound Mg_2Si. A typical alloy has the composition 0.7% magnesium, 1.0% silicon and 0.6% manganese. This alloy is not as strong as the duralumin but has greater ductility. It is used for ladders, scaffold tubes, container bodies, structural members for road and rail vehicles. The heat treatment is solution treatment at 510°C with precipitation hardening by quenching followed by precipitation hardening of 10 hours at about 165°C. Another group of alloys is based on aluminium-zinc-magnesium-copper, e.g. 5.5% zinc, 2.8% magnesium, 0.45% copper, 0.5% manganese. These alloys have the highest strength of the aluminium alloys and are used for structural applications in aircraft and spacecraft.

An alloy for use in the casting process must flow readily to all parts of the mould and on solidifying it should not shrink too much and any shrinking should not result in fractures. In choosing an alloy for casting the type of casting process being used needs to be taken into account. In sand casting the mould is made of sand bonded with clay or a resin. The cooling rate with such a method is relatively slow. A material for use by this method must give a suitable strength material after a slow cooling process. With die casting the mould is made of metal and the hot metal is injected into the die under pressure. This results in a fast cooling. A material for use by this method must develop suitable strength after fast cooling.

A family of aluminium alloys that can be used in the 'as cast' condition, i.e. no heat treatment is used, has aluminium with between 9 and 13% silicon. These alloys can be used for both sand and die casting. *Figure 4.5* shows the thermal equilibrium diagram

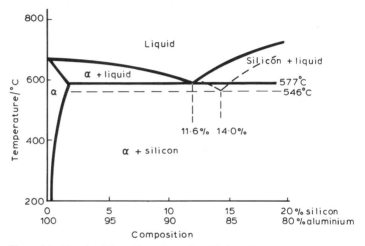

Figure 4.5   The aluminium-silicon thermal equilibrium diagram

for aluminium-silicon alloys. The addition of silicon to aluminium increases its fluidity, between about 9 to 13% giving a suitable fluidity for casting purposes. The eutectic for aluminium-silicon alloys has a composition of 11.6% silicon. An alloy of this composition changes from the liquid to the solid state without any change in temperature, alloys close to this composition solidify over a small temperature range (*Figure 4.6*). This makes them particularly suitable for die casting where a quick change from liquid to solid is required in order that a rapid ejection from the die can occur and high output rates achieved.

For the eutectic composition the microstructure shows a rather coarse eutectic structure of α phase and silicon. For an alloy having more silicon than the 11.6% eutectic value, the microstructure consists of silicon crystals in eutectic structure (*Figure 4.7*). The coarse eutectic structure, together with the presence of the embrittling silicon crystals, results is rather poor mechanical properties for the casting. The structure can however be made finer and the silicon crystal formation prevented by a process known as *modification*. This involves adding about 0.005 to 0.15% metallic sodium to the liquid alloy before casting. This produces a considerable refinement of the eutectic structure and also causes the eutectic composition to change to about 14.0% silicon. This displacement of the eutectic point is indicated on *Figure 4.5* by the dashed line. Thus for a silicon content below 14%, the structure, as modified, has α phase crystals

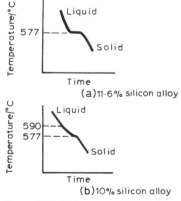

Figure 4.6   Cooling curves, (a) 11.6% silicon alloy, (b) 10% silicon alloy

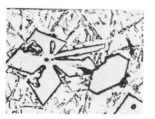

Figure 4.7   A 13% cast silicon alloy showing silicon with eutectic. (From Rollason E. C., *Metallurgy for Engineers*, Edward Arnold)

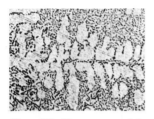

Figure 4.8   The same material as in *Figure 4.7* but modified by sodium showing aluminium with a fine eutectic (From Rollason, E. C., *Metallurgy for Engineers,* Edward Arnold)

in a finer eutectic structure (*Figure 4.8*). The result is an increase in both tensile strength and ductility and so a much better casting material.

The aluminium-silicon alloy is widely used for both sand and die casting, being used for many castings in cars, e.g. sumps, gear boxes and radiators. It is also used for pump parts, motor housings and a wide variety of thin walled and complex castings.

Other cast alloys that are not heat treated are aluminium-silicon-copper alloys, e.g. 5.0% silicon and 3.0% copper, and aluminium-magnesium-manganese alloys, e.g. 4.5% magnesium and 0.5% manganese. The silicon-copper alloys can be both sand and die cast, the magnesium-manganese alloys are however only suitable for sand casting. They have excellent corrosion resistance and are often used in marine environments.

The addition of copper, magnesium and other elements to aluminium alloys, either singly or in some suitable combination, can enable the alloy to be heat treated. Thus an alloy having 5.5% silicon and 0.6% magnesium can be subjected to solution treatment followed by precipitation hardening to give a high-strength casting material. Another heat treatable casting alloy has 4.0% copper, 2.0% nickel and 1.5% magnesium.

Example

Which of the following aluminium wrought alloys would be most suitable for use in a marine environment? (a) 1.25% manganese (b) 5.0% magnesium (c) 1.2% magnesium.

The aluminium alloy with the highest percentage magnesium (b) has the greatest corrosion resistance and so is most suitable for a marine environment.

Example

What is the effect on the mechanical properties of an aluminium – 12% silicon casting alloy of adding a very small amount of metallic sodium to the liquid alloy before casting?

This adding of sodium to the liquid alloy is known as modification and results in a refinement of the eutectic structure as well as a displacement of the eutectic point to 14% silicon. The effect on the mechanical properties is to increase the tensile strength and the ductility.

PROPERTIES OF ALUMINIUM ALLOYS

The table shows the properties that might be obtained with aluminium alloys. Page 33 explains the terms used in the table.

Composition	Condition	Tensile strength / N mm⁻²	Elongation / %	Hardness / HB
Wrought, non-heat treated alloys				
1.25% Mn	Annealed	110	30	30
	Hard	180	3	50
2.25% Mg	Annealed	180	22	45
	¼ hard	250	4	70
5.0% Mg	Annealed	300	16	65
	¼ hard	340	8	80
Wrought, heated treated alloys				
4.0% Cu, 0.8% Mg, 0.5% Si, 0.7% Mn	Annealed	180	20	45
	Solution treated, precipitation hardened	430	20	100

Composition	Condition	Tensile strength / N mm^{-2}	Elongation / %	Hardness / HB
4.3% Cu, 0.6% Mg, 0.8% Si, 0.75% Mn	Annealed	190	12	45
	Solution treated, precipitation hardened	450	10	125
0.7% Mg, 1.0% Si, 0.6% Mn	Annealed	120	15	47
	Solution treated, precipitation hardened	300	12	100
5.5% Zn, 2.8% Mg, 0.45% Cu, 0.5% Mn	Solution treated, precipitation hardened	500	6	170
Cast, non-heat treated alloys				
12% Si	Sand cast	160	5	55
	Die cast	185	7	60
5% Si, 3% Cu	Sand cast	150	2	70
	Die cast	170	3	80
4.5% Mg, 0.5% Mn	Sand cast	140	3	60
Cast, heat treated alloys				
5.5% Si, 0.6% Mg	Sand cast, solution treated, precipitation hardened	235	2	85
4.0% Cu, 2% Ni, 1.5% Mg	Sand cast, solution treated, precipitation hardened	275	1	110

The effects of the various alloying elements used with aluminium can be summarised as:

Copper	Increases strength. Precipitation heat treatment possible. Improves machineability
Manganese	Improves ductility. Improves, in combination with iron, the castability
Magnesium	Improves strength. Precipitation heat treatment possible with more than about 6%. Improves the corrosion resistance
Silicon	Improves castability, giving an excellent casting alloy. Improves corrosion resistance
Zinc	Lowers castability. Improves strength when combined with other alloying elements

COPPER

Copper has a density of 8.93 10^3 kg m^{-3}. It has very high electrical and thermal conductivity and can be manipulated readily by either hot or cold working. Pure copper is very ductile and relatively weak. The tensile strength and hardness can be increased by working, this does however decrease the ductility. Copper has good corrosion resistance. This is because there is a surface reaction between copper and the oxygen in the air which results in the formation of a thin protective oxide layer.

Very pure copper can be produced by an electrolytic refining process. An impure slab of copper is used as the anode while a pure thin sheet of copper is used as the cathode. The two electrodes are suspended in a warm solution of dilute sulphuric acid (*Figure 4.9*). The passage of an electric current through the arrangement causes copper to leave the anode and become deposited on the cathode. The result is a thicker, pure copper cathode, while the anode effectively

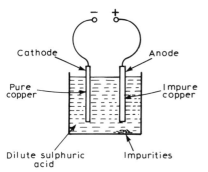

Figure 4.9 Basic arrangement for the electrolytic refining of copper

disappears, the impurities having fallen to the bottom of the container. The copper produced by this process is often called *cathode copper* and has a purity greater than 99.99%. It is used mainly as the raw material for the production of alloys, though there is some use as a casting material.

Electrolytic tough pitch high conductivity copper is produced from cathode copper which has been melted and cast into billets, and other suitable shapes, for working. It contains a small amount of oxygen, present in the form of cuprous oxide, which has little effect on the electrical conductivity of the copper. This type of copper should not be heated in an atmosphere where it can combine with hydrogen because the hydrogen can diffuse into the metal and combine with the cuprous oxide to generate steam. This steam can exert sufficient pressure to cause cracking of the copper.

Fire refined tough pitch high conductivity copper is produced from impure copper. In the fire refining process, the impure copper is melted in an oxidising atmosphere. The impurities react with the oxygen to give a slag which is removed. The remaining oxygen is partially removed by poles of green hardwood being thrust into the liquid metal, the resulting combustion removes oxygen from the metal. The resulting copper has an electrical conductivity almost as good as the electrolytic tough pitch high conductivity copper.

Oxygen-free high conductivity copper can be produced if, when cathode copper is melted and cast into billets, there is no oxygen present in the atmosphere. Such copper can be used in atmospheres where hydrogen is present.

Another method of producing oxygen-free copper is to add phosphorus during the refining. The effect of small amounts of phosphorus in the copper is a very marked decrease in the electrical conductivity, of the order of 20%. Such copper is known as *phosphorus deoxidised copper* and it can give good welds, unlike the other forms of copper.

The addition of about 0.5% arsenic to copper increases its tensile strength, especially at temperatures of about 400°C. It also improves its corrosion resistance but greatly reduces the electrical and thermal conductivities. This type of copper is known as *arsenical copper*.

Electrolytic tough pitch, high-conductivity copper finds use in high-grade electrical applications, e.g. wiring and busbars. Fire-refined, tough pitch, high-conductivity copper is used for standard electrical applications. Tough pitch copper is also used for heat exchangers and chemical plant. Oxygen-free, high-conductivity copper is used for high-conductivity applications where hydrogen may be present, electronic components and as the anodes in the electrolytic refining of copper. Phosphorus deoxidised copper is used in chemical plant where good weldability is necessary and for plumbing and general pipework. Arsenical copper is used for general engineering work, being useful to temperatures of the order of 400°C.

PROPERTIES OF COPPER   The table shows the properties that might be obtained with the various forms of copper. Page 33 explains the terms used in the table.

Composition %	Condition	Tensile strength /N mm⁻²	Elongation %	Hardness /HB
Electrolytic tough-pitch, high-conductivity copper				
99.90 min	Annealed	220	50	45
0.05 oxygen				
	Hard	400	4	115
Fire refined tough-pitch, high conductivity copper				
99.85 min	Annealed	220	50	45
0.05 oxygen				
	Hard	400	4	115
Oxygen-free high-conductivity copper				
99.95 min	Annealed	220	60	45
	Hard	400	6	115
Phosphorus deoxidised copper				
99.85 min	Annealed	220	60	45
0.013–0.05 P				
	Hard	400	4	115
Arsenical copper				
99.20 min	Annealed	220	50	45
0.05 oxygen,				
0.3–0.5 As				
	Hard	400	4	115

The following information from a manufacturer's information sheet describes some of the other forms of copper (Courtesy of McKechnie Metals Ltd.).

Copper in its purer forms is not the easiest of metals to machine and two alloys are available which, while considerably easier to machine, yielding well-broken swarf and good surface finish, are only slightly lower in conductivity and ductility.

The addition of 0.5% tellurium forms a chip-breaking compound, copper telluride, but still enables a conductivity of 90% to be maintained. The addition of 0.4% sulphur produces a similar effect through the formation of copper sulphide and provides a less expensive machinable grade of copper with a conductivity of 95%.

Tellurium copper (phosphorus deoxidised) 99.5% copper, trace phosphorus, 0.5% tellurium. Mechanical properties drawn: tensile strength 300 N mm⁻², elongation 15%, hardness 100 HB. Fabrication properties: machinability rating excellent, cold working fair, hot working moderate, hot working range 800–850°C.

Sulphur copper: 99.6% copper, 0.4% sulphur. Mechanical properties drawn: tensile strength 250 N mm⁻², elongation 20%, hardness 90 HB. Fabrication properties, machinability rating good, cold working fair, hot working moderate, hot working range 800–850°C.

COPPER ALLOYS The most common elements with which copper is alloyed are zinc, tin, aluminium and nickel. The copper-zinc alloys are referred to as brasses, the copper-tin alloys as tin bronzes, the copper-aluminium alloys as aluminium bronzes and the copper-nickel alloys as cupronickels, though where zinc is also present they are called nickel silvers. A less common copper alloy involves copper and beryllium.

The copper-nickel thermal equilibrium diagram is rather simple as the two metals are completely soluble in each other in both the liquid and solid states. The copper-zinc, copper-tin and copper-aluminium thermal equilibrium diagrams are however rather complex. In all

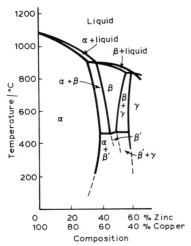

Figure 4.10 Thermal equilibrium diagram for copper-zinc alloys

cases, the α phase solid solutions have the same types of microstructure and are ductile and suitable for cold working. When the amount of zinc, tin or aluminium exceeds that required to saturate the α solid solution a β phase is produced. The microstructures of these β phases are similar and alloys containing this phase are stronger and less ductile. They cannot be readily cold worked and are hot worked or cast. Further additions lead to yet further phases which are hard and brittle.

The *brasses* are copper-zinc alloys containing up to about 43% zinc. *Figure 4.10* shows the relevant part of the thermal equilibrium diagram. Brasses with between 0 and 35% zinc solidify as α solid solutions, usually cored (*Figure 4.12*). These brasses have high ductility and can readily be cold worked. *Gilding brass,* 15% zinc, is used for jewellery because it has a colour resembling that of gold and can so easily be worked. *Cartridge brass,* 30% zinc and frequently referred to as *70/30 brass,* is used where high ductility is required with relatively high strength. It is called cartridge brass because of its use in the production of cartridge and shell cases. The brasses in the 0 to 30% zinc range all have their tensile strength and hardness increased by working, but the ductilities decrease.

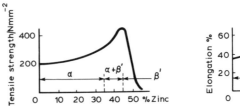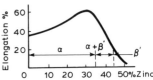

Figure 4.11 Strength and ductility for copper-zinc alloys

Brasses with between 35 and 46% zinc solidify as a mixture of two phases (*Figure 4.13*). Between about 900°C and 453°C the two phases are α and β. At 453°C this β phase transforms to a low temperature modification referred to as β' phase. Thus at room temperature the two phases present are α and β'. The presence of the β' phase produces a drop in ductility but an increase in tensile strength to the maximum value for a brass (*Figure 4.11*). These brasses are known as *alpha-beta or duplex brasses.* These brasses are not cold worked but have good properties for hot forming processes,

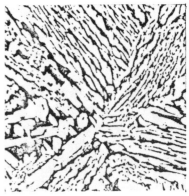

Figure 4.12 A 70% copper – 30% zinc alloy. The structure is heavily cored

Figure 4.13 A 60% copper – 40% zinc alloy. This consists of light α phase in a matrix of dark β'

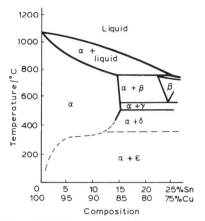

Figure 4.14 Thermal equilibrium diagram for tin bronzes

e.g. extrusion. This is because the β phase is more ductile than the β' phase and hence the combination of α plus β gives a very ductile material. The hot working should take place at temperatures in excess of 453°C. The name *Muntz metal* is given to a brass with 60% copper – 40% zinc.

The addition of lead to Muntz metal improves considerably the machining properties, without significantly changing the strength and ductility. *Leaded Muntz metal* has 60% copper, 0.3 to 0.8% lead and the remainder zinc.

Copper-zinc alloys containing just the β' phase have little industrial application. The presence of γ phase in a brass results in a considerable drop in strength and ductility, a weak brittle product being obtained.

Copper-tin alloys are known as *tin bronzes. Figure 4.14* shows the thermal equilibrium diagram for such alloys. The dashed lines on the diagram indicate the phase that can occur with extremely slow cooling. The structure that normally occurs with up to about 10% tin are predominantly α solid solutions. Higher percentage tin alloys will invariably include a significant amount of the δ phase. This is a brittle intermetallic compound, the α phase being ductile.

Bronzes that contain up to about 8% tin are α bronzes and can be cold worked. In making bronze, oxygen can react with the metals and lead to a weak alloy. Phosphorus is normally added to the liquid metals to act as a deoxidiser. Some of the phosphorus remains in the final alloy. This type of alloy is known as *phosphor bronze*. These alloys are used for springs, bellows, electrical contacts, clips, instrument components. A typical phosphor bronze might have about 95% copper, 5% tin and 0.02 to 0.40% phosphorus.

The above discussion refers to wrought phosphor bronzes. Cast phosphor bronzes contain between 5 and 13% tin with as much as 0.5% phosphorus. A typical cast phosphor bronze used for the production of bearings and high grade gears has about 90% copper, 10% tin and a maximum of 0.5% phosphorus. This material is particularly useful for bearing surfaces; it has a low coefficient of friction and can withstand heavy loads. The hardness of the material occurs by virtue of the presence of both δ phase and a copper-phosphorus compound.

Casting bronzes that contain zinc are called *gunmetals.* This reduces the cost of the alloy and also makes unnecessary the use of phosphorus for deoxidation as this function is performed by the zinc. *Admiralty gunmetal* contains 88% copper, 10% tin and 2% zinc. This alloy finds general use for marine components, hence the word 'Admiralty'.

Copper-aluminium alloys are known as *aluminium bronzes. Figure 4.15* shows the thermal equilibrium diagram for such alloys. Up to about 9% aluminium gives *alpha bronzes,* such alloys containing just the α phase. Alloys with up to about 7% aluminium can be cold worked readily. *Duplex alloys* with about 10% aluminium are used for casting. Aluminium bronzes have high strength, good resistance to corrosion and wear. These corrosion and wear properties arise because of the thin film of aluminium oxide formed on the surfaces. Typical applications of such materials are high-strength and highly corrosion-resistant items in marine and chemical environments, e.g. pump casings, gears, valve parts.

Alloys of copper and nickel are known as *cupronickels,* though if

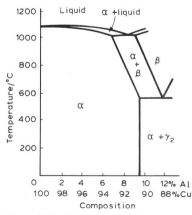

Figure 4.15 Thermal equilibrium diagram for aluminium bronzes

zinc is also present they are referred to as *nickel silvers. Figure 1.18* shows the thermal equilibrium diagram for copper-nickel alloys. Copper and nickel are soluble in each other in both the liquid and solid states, they thus form a solid solution whatever the proportions of the two elements. They are thus α phase over the entire range and suitable for both hot and cold working over the entire range. The alloys have high strength and ductility, and good corrosion resistance. The 'silver' coinage in use in Britain is a 75% copper –25% nickel alloy. The addition of 1 to 2% iron to the alloys increases their corrosion resistance. A 66% copper – 30% nickel – 2% manganese –2% iron alloy is particularly resistant to corrosion, and erosion, and is used for components immersed in moving sea water.

Nickel silvers have a silvery appearance and find use for items such as knives, forks and spoons. The alloys can be cold worked and usually contain about 20% nickel, 60% copper and 20% zinc.

Copper alloyed with small percentages of beryllium can be precipitation heat treated to give alloys with very high tensile strengths, such alloys being known as *beryllium bronzes,* or *beryllium copper.* The alloys are used for high-conductivity, high-strength electrical components, springs, clips and fastenings.

Example
Two brasses, an alpha and a duplex brass, are available. Which brass should be selected for cold working?
The alpha brass.

Example
Basis brass has the composition 63% copper – 37% zinc. What type of mechanical properties might be expected of such a brass?
Brasses with between 0 and about 35% zinc solidify as alpha brasses. Basis brass is thus likely to be predominantly alpha but with some beta phase. Because of the predominant amount of alpha brass the brass is likely to be reasonably ductile and capable of being cold worked. The presence of a small amount of beta phase will result in an increase in tensile strength. The brass is likely therefore to be a reasonable compromise, having a relatively high tensile strength and ductility. In fact, basis brass is a general purpose alloy, being widely used for general hardware.

PROPERTIES OF COPPER ALLOYS

The table shows the properties that might be obtained with copper alloys. Page 33 explains the terms used in the table.

Composition %	Condition	Tensile strength /N mm^{-2}	Elongation %	Hardness /HB
Brasses				
90 copper, 10 zinc	Annealed	280	48	65
80 copper, 20 zinc	Annealed	320	50	67
70 copper, 30 zinc	Annealed	330	70	65
	Hard	690	5	185
60 copper, 40 zinc	Annealed	380	40	75

Composition %	Condition	Tensile strength /N mm⁻²	Elongation %	Hardness /HB
Tin bronzes				
95 copper,	Annealed	340	55	80
5 tin, 0.02–0.40	Hard	700	6	200
phosphorus				
91 copper,	Annealed	420	65	90
8–9 tin,	Hard	850	4	250
0.02–0.40				
phosphorus				
Gunmetal				
88 copper,	Sand cast	300	20	80
10 tin,				
2 zinc				
Aluminium bronzes				
95 copper,	Annealed	370	65	90
5 aluminium	Hard	650	15	190
88 copper,	Sand cast	545	30	110
9.5 aluminium,				
2.5 iron				
Cupronickels				
87.5 copper, 10	Annealed	320	40	155
nickel, 1.5 iron,				
1 manganese				
75 copper, 25	Annealed	360	40	90
nickel, 0.5	Hard	600	5	170
manganese				
Nickel silver				
64 copper, 21	Annealed	400	50	100
zinc, 15 nickel	Hard	600	10	180
Beryllium bronzes				
98 copper, 1.7	Solution	1200	3	370
beryllium, 0.2 to	treated,			
0.6 cobalt and	precipitation			
nickel	hardened			

There are a considerable number of copper alloys the following indicates the type of selection that could be made.

Electrical conductors	Electrolytic tough-pitch, high-conductivity copper
Tubing and heat exchangers	Phosphorus deoxidised copper is generally used. Muntz metal, cupronickels or naval brass (62% copper–37% zinc–1% tin) is used if the water velocities are high
Pressure vessels	Phosphorus deoxidised copper, copper-clad steel or aluminium bronze
Bearings	Phosphor bronze. Other bronzes and brasses with some lead content are used in some circumstances
Gears	Phosphor bronze. For light duty gunmetals, aluminium bronze or die cast brasses may be used
Valves	Aluminium bronze
Springs	Phosphor bronze, nickel silver, basis brass are used for low cost springs. Beryllium bronze is the best material

MAGNESIUM   Magnesium has a density of $1.7\ 10^3\ kg\ m^{-3}$ and thus a very low density compared with other metals. It has an electrical conductivity of about 60% of that of copper, as well as a high thermal conductivity. It has a low tensile strength, needing to be alloyed with other metals to improve its strength. Under ordinary atmospheric conditions magnesium has good corrosion resistance, which is provided by an oxide layer that develops on the surface of the magnesium in air. However, this oxide layer is not completely impervious, particularly in air that contains salts, and thus the corrosion resistance can be low under adverse conditions. Magnesium is only used generally in its alloy form, the pure metal finding little application.

MAGNESIUM ALLOYS

Because of the low density of magnesium, the magnesium-base alloys have low densities. Thus magnesium alloys are used in applications where lightness is the primary consideration, e.g. in aircraft and spacecraft. Aluminium alloys have higher densities than magnesium alloys but can have greater strength. The strength-to-weight ratio for magnesium alloys is however greater than that for aluminium alloys. Magnesium alloys also have the advantage of good machinability and weld readily.

Magnesium-aluminium-zinc alloys and magnesium-zinc-zirconium are the main two groups of alloys in general use. Small amounts of other elements are also present in these alloys. The composition of an alloy depends on whether it is to be used for casting or working, i.e. a wrought alloy. The cast alloys can often be heat treated to improve their properties.

A general-purpose wrought alloy has about 93% magnesium – 6% aluminium – 1% zinc – 0.3% manganese. This alloy can be forged, extruded and welded, and has excellent machinability. A high strength wrought alloy has 96.4% magnesium – 3% zinc – 0.6% zirconium. A general-purpose casting alloy has about 91% magnesium – 8% aluminium – 0.5% zinc – 0.3% manganese. A high-strength casting alloy has 94.8% magnesium – 4.5% zinc –0.7% zirconium. Both these casting alloys can be heat treated.

PROPERTIES OF MAGNESIUM ALLOYS

The table shows the properties that might be obtained with magnesium alloys. Page 33 explains the terms used in the table.

Composition %	Condition	Tensile strength /N mm^{-2}	Elongation %	Hardness /HB
Wrought alloys				
93 magnesium, 6	Forged	290	8	65
aluminium, 1 zinc,	Extruded	310	8	70
0.3 manganese				
96.4 magnesium,				
3 zinc, 0.6				
zirconium				
Cast alloys				
91 magnesium, 8	As cast	140	2	55
aluminium, 0.5	Heat treated	200	6	75
zinc, 0.6 zirconium	Heat treated	230	5	70
94.8 magnesium,				
4.5 zinc, 0.7				
zirconium				

NICKEL

Nickel has a density of 8.88 10^3 kg m^{-3} and a melting point of 1455°C. It possesses excellent corrosion resistance, hence it is used often as a cladding on a steel base. This combination allows the corrosion resistance of the nickel to be realised without the high cost involved in using entirely nickel. Nickel has good tensile strength and maintains it at quite elevated temperatures. Nickel can be both cold and hot worked, has good machining properties and can be joined by welding, brazing and soldering.

Nickel is used in the food processing industry in chemical plant, and in the petroleum industry, because of its corrosion resistance and strength. It is also used in the production of chromium-plated mild steel, the nickel forming an intermediate layer between the steel and the chromium. The nickel is electroplated on to the steel.

NICKEL ALLOYS

Nickel is used as the base metal for a number of alloys with excellent corrosion resistance and strength at high temperatures. One group of alloys is based on nickel combined with copper, the thermal equilibrium diagram for these alloys being on page 9. A common nickel-copper alloy is known as *Monel* which has 68% nickel, 30% copper and 2% iron. It is highly resistant to sea water, alkalis, many acids and superheated steam. It has also high strength, hence its use for steam turbine blades, food processing equipment and chemical engineering plant components.

Another common nickel alloy is known as *Inconel* which contains 78% nickel, 15% chromium and 7% iron. The alloy has a high strength and excellent resistance to corrosion at both normal and high temperatures. It is used in chemical plant, aero-engines, as sheaths for electric cooker elements, steam turbine parts and heat treatment equipment.

The *Nimonic* series of alloys are basically nickel–chromium alloys, essentially about 80% nickel and 20% chromium. They have high strength at high temperatures and are used in gas turbines for discs and blades.

PROPERTIES OF NICKEL ALLOYS

The table shows the properties that might be obtained with nickel alloys. Page 33 explains the terms used in the table.

Composition %	Condition	Tensile strength /N mm^{-2}	Elongation %	Hardness /HB
68 nickel, 30 copper, 2 iron	Annealed	500	40	110
	Cold worked	840	8	240
78 nickel, 15 chromium, 7 iron	Annealed	700	35	170
	Cold worked	1050	15	290

TITANIUM

Titanium has a relatively low density, $4.5 \, 10^3 \, kg \, m^{-3}$, just over half that of steel. It has a relatively low tensile strength when pure but alloying gives a considerable increase in strength. Because of the low density of titanium its alloys have a high strength-to-weight ratio. Also, it has excellent corrosion resistance. However, titanium is an expensive metal, its high cost reflecting the difficulties experienced in the extraction and forming of the material; the ores are quite plentiful.

TITANIUM ALLOYS

The main alloying elements used with titanium are aluminium, copper, manganese, molybdenum, tin, vanadium and zirconium. A ductile, heat treatable, alloy has 97.5% titanium – 2.5% copper and it can be welded and formed. A higher-strength alloy, which can also be welded and formed, has 92.5% titanium – 5% aluminium – 2.5% tin. A very high strength alloy has 82.5% titanium – 11% tin – 4% molybdenum – 2.25% aluminium – 0.25% silicon. This alloy can be heat-treated and forged.

The titanium alloys all show excellent corrosion resistance, have good strength-to-weight ratios, can have high strengths, and have good properties at high temperatures. They are used for compressor blades, engine forgings, components in chemical plant, and other duties where their properties make them one of the few possible choices despite their high cost.

PROPERTIES OF TITANIUM ALLOYS

The table shows the properties that might be obtained with titanium alloys. Page 33 explains the terms used in the table.

Composition %	Condition	Tensile strength /N mm^{-2}	Elongation %	Hardness /HB
97.5 titanium, 2.5 copper	Heat treated	740	15	360
92.5 titanium, 5 aluminium, 2.5 tin	Annealed	880	16	360
82.5 titanium, 11 tin, 4 molybdenum, 2.25 aluminium, 0.25 silicon	Heat treated	1300	15	380

ZINC

Zinc has a density of 7.1×10^3 kg m^{-3}. Pure zinc has a melting point of only 419°C and is a relatively weak metal. It has good corrosion resistance, due to the formation of an impervious oxide layer on the surface. Zinc is frequently used as a coating on steel in order to protect that material against corrosion, the product being known as galvanised steel.

ZINC ALLOYS

The main use of zinc alloys is for die-casting (see page 49). They are excellent for this purpose by virtue of their low melting points and the lack of corrosion of dies used with them. The two alloys in common use for this purpose are known as alloy A and alloy B. *Alloy A,* the most used of the two, has the composition of 3.9 to 4.3% (max) aluminium, 0.03% (max) copper, 0.03 to 0.06% (max) magnesium, the remainder being zinc. *Alloy B* has the composition 3.9 to 4.3% (max) aluminium, 0.75 to 1.25% (max) copper, 0.03 to 0.06% (max) magnesium, with the remainder being zinc. Alloy A is the more ductile, alloy B has the greater strength.

The zinc used in the alloys has to be extremely pure so that little, if any, other impurities are introduced into the alloys, typically the required purity is 99.99%. The reason for this purity is that the presence of very small amounts of cadmium, lead or tin renders the alloy susceptible to intercrystalline corrosion. The products of this corrosion cause a casting to swell and may lead to failure in service.

After casting, the alloys undergo a shrinkage which takes about a month to complete; after that there is a slight expansion. A casting can be *stabilised* by annealing at 100°C for about 6 hours.

Zinc alloys can be machined and, to a limited extent, worked. Soldering and welding is not generally feasible.

Zinc alloy die-castings are widely used in domestic appliances, for toys, car parts such as door handles and fuel pump bodies, optical instrument cases.

The following extract from a manufacturer's information sheet illustrates some of the arguments for zinc and die-casting. (Courtesy of Dynacast International Ltd).

The case for zinc

The most widely used alloy in die-casting.

Low melting-point and excellent fluidity.

Easiest metal to cast.

Best alloy for small parts of complex shape and thin wall section.

Easily electroplated and mechanically polished.

Best mechanical properties after brass, which has poor castability.

The case for die-casting

Fastest of all casting processes.

High speed, economical production.

Particularly suitable for small parts.

High-pressure die-casting results in fine grain structure and good mechanical properties.

Complex, close tolerance parts made relatively easily.

PROPERTIES OF ZINC ALLOYS

The table shows the properties that might be obtained with zinc die-casting alloys. Page 33 explains the terms used in the table.

Composition %	Condition	Tensile strength /N mm^{-2}	Elongation %	Hardness /HB
Alloy A	As cast	285	10	83
Alloy B	As cast	330	7	92

COMPARISON OF NON-FERROUS ALLOYS

The following table shows the range of strengths that are obtained with the non-ferrous alloys considered in this chapter.

Alloys	Tensile strength/N mm^{-2}
Aluminium	100 to 550
Copper	200 to 1300
Magnesium	150 to 350
Nickel	400 to 1300
Titanium	400 to 1600
Zinc	200 to 350

Thus, if the main consideration in the choice of an alloy is that it must have high strength, then the choice, if limited to the alloys listed above, would be titanium, with the second choice nickel or copper alloys.

In some applications it is not just the strength of a material that is important but the strength/weight ratio. This is particularly the case in aircraft or spacecraft where not only is strength required but also a low mass. The following table shows the densities of the alloys and the tensile strength/density ratio.

Alloys	Density /10^3 kg m^{-3}	Tensile strength/density /N mm^{-2} ÷ 10^3 kg m^{-3}
Aluminium	2.7	37 to 200
Copper	8	25 to 160
Magnesium	1.8	110 to 190
Nickel	8.9	47 to 146
Titanium	4.5	89 to 356
Zinc	6.7	30 to 52

Titanium gives the best possible strength/weight ratio. Magnesium, with its relatively low strength, has however a fairly high strength/weight ratio by virtue of its very low density. Aluminium, magnesium and titanium alloys are widely used in aircraft.

Another consideration that may affect the choice of an alloy is its corrosion resistance. In general the most resistant are titanium alloys, with the least resistant being magnesium. The rough order of descending corrosion resistance is: titanium, copper, nickel, zinc, aluminium and magnesium.

A vital factor in considering the choice of a material is the cost. The following table gives the relative costs of the alloys, the costs all being relative to aluminium. The costs are given in terms of the cost per unit mass.

Alloy	Relative cost per unit mass
Aluminium	1
Copper	1 to 2
Magnesium	3
Nickel	5
Titanium	25
Zinc	0.5

Titanium is the most expensive of the alloys, zinc the least expensive. However in practice the use of a high-strength material may mean that a thinner section can be used and so less mass of alloy is required.

The process to be used to shape the material is another constraint on the selection of the material. With ductile materials, the drag forces on a cutting tool are high and the swarf tends to build up. These lead to poor machinability. Annealed aluminium has a high ductility and thus has poor machineability. A half-hard aluminium alloy can however have a good machineability, the ductility being much less. Materials can have their machineability improved by introducing other materials, such as lead, into the alloy concerned. The introduced material is in the form of particles which help to break up the swarf into small chips. Very hard materials may present machining problems in that the cutting tool needs to be harder than the material being machined.

Example
Which type of alloy would be the optimum choice for a new high-speed aircraft where high strength at high temperatures combined with a low density is required?

Titanium alloys give high strengths at high temperatures and have a low density.

Example
Which type of alloy combines the following properties: can be die-cast, reasonable corrosion resistance and cheap. Strength is not particularly required.

Zinc alloys can be die-cast and are cheap. They also have reasonable corrosion resistance.

PROBLEMS
1. What is the effect on the strength and ductility of aluminium of (a) the purity of the aluminium, (b) the temper?

2. Describe the effect on the strength of aluminium of the percentage of magnesium alloyed with it.

3. Describe the solution treatment and precipitation hardening processes for aluminium-copper alloys.

4. Describe the features of aluminium-silicon alloys which makes them suitable for use with die-casting.

5. What is the effect of heat treatment on the properties of an aluminium alloy, such as a 4.3% copper – 0.6% magnesium – 0.8% silicon – 0.75% manganese?

6. Ladders are often made from an aluminium alloy. What are the properties required of that material in this particular use?

7. Explain with the aid of thermal equilibrium diagram how the addition of a small amount of sodium to the melt of an aluminium–silicon alloy changes the properties of alloys with, say, 12% silicon.

8. What are the differences between the following forms of copper: electrolytic tough-pitch, high-conductivity copper; fire-refined, tough-pitch, high-conductivity copper; oxygen-free, high-conductivity copper; phosphorus-deoxidised copper; arsenical copper.

9. Which form of copper should be used in an atmosphere containing hydrogen?

10. What is the effect of cold work on the properties of copper?

11. What is the effect of the percentage of zinc in a copper-zinc alloy on its strength and ductility?

12. What brass composition would be most suitable for applications requiring (a) maximum tensile strength, (b) maximum ductility, (c) the best combined tensile strength and ductility?

13. In general, what are the differences in (a) composition, (b) properties of α phase and duplex alloys of copper?

14. The 'silver' coinage used in Britain is made from a cupro-nickel alloy. What are the properties required of this material for such a use?

15. What are the constituent elements in (a) brasses, (b) phosphor bronzes and (c) cupro-nickels?

16. The name Muntz metal is given to a 60% copper – 40% zinc alloy. Some forms of this alloy also include a small percentage of lead. What is the reason for the lead?

17. What percentage of aluminium would be likely to be present in an aluminium bronze that is to be cold worked?

18. What is meant by the term strength-to-weight ratio? Magnesium alloys have a high strength-to-weight ratio; of what significance is this in the uses to which the alloys of magnesium are put?

19. How does the corrosion resistance of magnesium alloys compare with other non-ferrous alloys?

20. What are the general characteristics of the nickel-copper alloys known as Monel?

21. Though titanium alloys are expensive compared with other non-ferrous alloys, they are used in modern aircraft such as *Concorde.* What advantages do such alloys possess which outweighs their cost?

22. What problems can arise when impurities are present in zinc die-casting alloys?

23. What is the purpose of 'stabilising annealing' for zinc die-casting alloys?

24. Describe the useful features of zinc die-casting alloys which makes them so widely used.

25. This question is based on the aluminium-silicon equilibrium diagram given in *Figure 4.5.*
    (a) At what temperature will the solidification of a 85% aluminium – 15% silicon start? At what temperature will it be completely solid?
    (b) The above alloy is modified by the addition of a small amount of sodium. At what temperature will solidification start? At what temperature will it be completely solid?

26. This question is based on the copper-zinc thermal equilibrium diagram given in *Figure 4.10.*
    (a) What is the temperature at which a 85% copper – 15% zinc alloy begins to solidify?
    (b) The pouring temperature used for casting is about 200°C above the liquidus temperature. What is the pouring temperature for the above alloy?

27. This question is based on the copper-zinc thermal equilibrium diagram given in *Figure 4.10.*
    (a) Which of the following brasses would you expect to be just α solid solution? (i) 10% zinc – 90% copper, (ii) 20% zinc – 80% copper, (iii) 40% zinc – 60% copper.
    (b) How would you expect the properties of the above brasses to differ? How are the differences related to the phases present in the alloys?

28. This question is based on the copper-nickel thermal equilibrium diagram given in *Figure 1.18.*
    (a) How does the copper-nickel thermal equilibrium diagram differ from that of copper-zinc?
    (b) Over what range of compositions will copper-nickel alloys be α solid solution?

29. The following group of questions is concerned with justifying the choice of a particular alloy for a specific application.
    (a) Why are zinc alloys used for die-casting?
    (b) Why are magnesium alloys used in aircraft?
    (c) Why are milk bottle caps made of an aluminium alloy?
    (d) Why are titanium alloys extensively used in high-speed aircraft?
    (e) Why are domestic water pipes made of copper?
    (f) Why is cartridge brass used for the production of cartridge cases?
    (g) Why are the ribs of hand gliders made of an aluminium alloy?
    (h) Why are nickel alloys used for gas-turbine blades?
    (i) Why is brass used for cylinder lock keys?
    (j) Why are kitchen pans made of aluminium or copper alloys?

(k) Why is copper used for gaskets?

(l) Why are electrical cables made of copper?

(m) Why are cupro-nickels used for tubes in desalination plants?

30. Compare the properties of ferrous and non-ferrous alloys, commenting on the relative ease of processing.

5 Polymers

Objectives: At the end of this chapter you should be able to:
Describe the basic structural features of thermoplastic, thermosetting and elastomer materials, relating the mechanical and thermal properties to the structures.
Describe the properties of common plastics, relating the properties to their structures.

POLYMER STRUCTURE

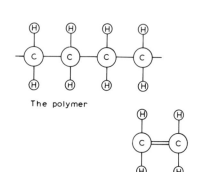

The polymer

The monomer

Figure 5.1 The polymer, polyethylene

The plastic washing-up bowl, the plastic measuring rule, the plastic cup – these are examples of polymeric materials. The molecules in these plastics are very large molecules. A molecule of oxygen consists of just two oxygen atoms joined together. A molecule in the plastic may have thousands of atoms all joined together in a long chain. The backbones of these long molecules are chains of carbon atoms. Carbon atoms are able to form strong bonds with themselves and produce long chains of carbon atoms to which other atoms can become attached.

The term *polymer* is used to indicate that a compound consists of many repeating structural units. The prefix 'poly' means many. Each structural unit in the compound is called a *monomer.* Thus the plastic polyethylene is a polymer which has as its monomer the substance ethylene. For many plastics the monomer can be determined by deleting the prefix 'poly' from the name of the polymer. *Figure 5.1* shows the basic form of a polymer.

If you apply heat to the plastic washing up bowl the material softens. Removal of the heat causes the material to harden again. Such a material is said to be *thermoplastic.* The term implies that the material becomes 'plastic' when heat is applied.

If you applied heat to a plastic cup you might well find that the material did not soften but charred and decomposed. Such a material is said to be a *thermosetting plastic.*

Another type of polymer is the elastomers. Rubber is an elastomer. An *elastomer* is a polymer which by its structure allows considerable extensions which are reversible.

The thermoplastic, thermosetting and elastomer materials can be distinguished by their behaviour when forces are applied to them to cause stretching. Thermoplastic materials are generally flexible and relatively soft; if heated they become softer and more flexible. Thermosetting materials are rigid and hard with little change with an increase in temperature. Elastomers can be stretched to many times their initial length and still spring back to their original length when released. These different types of behaviours of polymers can be explained in terms of differences in the ways the long molecular chains are arranged inside the material.

Figure 5.2 shows some of the forms the molecular chains can take. These forms can be described as linear, branched and cross-linked. The linear chains have no side branches or chains or cross links with

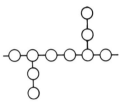

Linear polymer chain
(a)

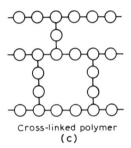

Branched polymer chain
(b)

Cross-linked polymer
(c)

Figure 5.2 (a) Linear polymer chain, (b) Branched polymer chain, (c) Cross-linked polymer

other chains. Linear chains can move readily past each other. If however the chain has branches there is a reduction in the ease with which chains can move past each other. This shows itself in the material being more rigid and having a higher strength. If there are cross-links a much more rigid material is produced in that the chains cannot slide past each other at all.

Polyethylene, a thermoplastic material, has linear molecular chains (in the high density version). Polyethylene is easily stretched and is not rigid. Because the chains are independent of each other they can easily flow past each other and so the material has a relatively low melting point, no energy being needed to break bonds between chains. The absence of bonds between chains also means that as none are broken when the material is heated the removal of heat allows the material to revert to its initial harder state.

Some thermoplastic materials have molecules with side branches; the effect of these is to give a harder and more rigid material. Polypropylene is such a material, being harder and more rigid than polyethylene. Another consequence of a material having branched chains is that, as they do not pack so readily together in the material as linear chains, the material will generally have a lower density than the linear chain material.

Thermosetting materials are cross-linked polymers and are rigid. As energy is needed to break bonds, before flow can occur, thermosetting materials have higher melting points than thermoplastic materials having linear or branched chains. Also the effect of heat is not reversible; when heat causes bonds to break an irreversible change to the structure of the material is produced. Bakelite is an example of a thermosetting material. It can withstand temperatures up to 200°C, but most thermoplastics are not used above 100°C.

Elastomers have linear molecular chains. In the material these chains are all tangled up and there is no order in the packing of the molecular chains in the material. These tangled chains give a relatively open structure with the large amount of empty space between the tangled chains. When forces are applied to the material the chains are able to move very easily within the voids. It is this which accounts for the very high extensions possible with elastomers.

CRYSTALLINITY IN POLYMERS

A crystalline structure is one in which there is an orderly arrangement of particles (see Chapter 1). A structure in which the arrangement is completely random is said to be *amorphous*. Many polymers are amorphous, the molecules in the material are completely randomly arranged. Highly cross-linked polymers (as in *Figure 5.2c*) are invariably armorphous. Linear polymers (*Figure 5.2a*) can be amorphous. *Figure 5.3* shows the type of structure that might occur for such a structure, the linear polymer chains being all tangled up.

Linear polymer chains can however assume an arrangement which is orderly. *Figure 5.4* shows the type of arrangement of chains that can occur, the linear chains folding backwards and forwards on themselves. The arrangement is said to be *crystalline*. The tendency of a polymer to crystallise is determined by the form of the polymer chains. Linear polymers can crystallise to quite an extent, complete crystallisation is not however obtained in that there are invariably some regions of disorder. Polyethylene can occur as linear chains and

Figure 5.3 A linear amorphous polymer. Individual atoms are not shown, the chains being represented by lines

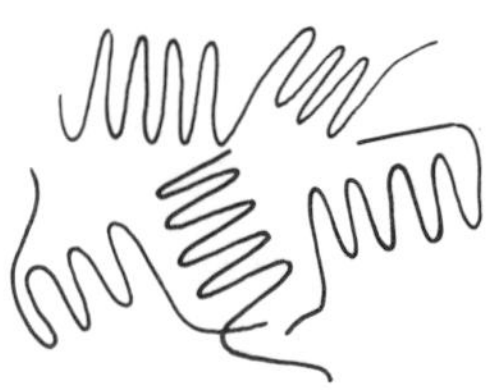

Figure 5.4 Folded linear polymer chains

in this form can have a crystallinity as high as 95%, i.e. of the entire piece of polyethylene, 95% of the material will have an orderly structure. Polymers with side chains show less tendency to crystallise. Thus the branched form of polyethylene may only give 50% crystallinity. Cross-linked polymers have zero crystallinity.

Crystallinity in polymers affects the properties of the polymers. Linear polyethylene with 95% crystallinity has a density of about 950 kg m^{-3} and a melting point of 135°C. Branched polyethylene with 50% crystallinity has a density of about 920 kg m^{-3} and a melting point of 115°C. The greater the crystallinity the greater the density, i.e. the more closely packed the molecules can be. The two forms of polyethylene are often known as high density and low density polyethylene. Also the greater the crystallinity the higher the melting point, i.e. the more closely packed the molecules, the stronger the forces between them and so the greater the energy that has to be supplied to separate the molecules and melt the polymer.

Crystallinity also lowers the solubility of polymers in solvents. It does however lead to stiffer, stronger materials (i.e. higher tensile modulus and tensile strength).

PLASTICS The term *plastics* is commonly used to describe materials based on polymers, i.e. long chain structures or networks. Such materials invariably contain other substances which are added to the polymers to give the required properties. Stabilisers, plasticisers and fillers are additives that are used. The term plastic is also restricted to those polymeric materials that are not elastomers.

Some plastics are damaged by ultraviolet radiation. Thus the effect of protracted periods of sunlight can lead to a deterioration of mechanical properties as well a reduction in transparency or change in colour. An ultraviolet absorber is thus often added to plastics, carbon black being often used. Such an additive is called a *stabiliser*.

The term *plasticiser* is used for the material added to a polymer to make it more flexible. In one form the plasticiser may be a liquid that is introduced after the polymer chains have been produced. The liquid disperses through the solid, filling up the spaces between the chains. The effect of this is the same as having a lubricant between two metal surfaces, so the polymer chains slide more easily past each other. One of the problems with such plasticised polymers is that the plasticiser can move out of the material with time. The vinyl upholstery used for car seats can lose its plasticiser and become more brittle. On a hot day the vinyl may feel greasy because of the plasticiser having migrated to the surface.

Another form of plasticisation involves replacing some of the links in the polymer chains by molecules which can more easily be deformed and so permit the chains to slide past each other more easily. Flexibility is thus increased.

The properties and the cost of a plastic can be markedly affected by the addition of other substances, these being termed *fillers*. The following table shows some of the common fillers used and their effects on the properties of the plastic.

Where the filler improves the tensile strength it generally does so by reducing the mobility of the polymer chains. An important consideration however in the use of any filler is the fact that the fillers are generally cheaper than the polymer and thus reduce the

Filler	Effect on properties
Asbestos	Improves temperature resistance, i.e. the plastic does not deform until a higher temperature is attained. Decreases strength and rigidity
Cotton flock	Increases impact strength but reduces electrical properties and water resistance
Cellulose fibres	Increases tensile strength and impact strength
Glass fibres	Increases tensile strength but lowers ductility. Makes the plastic stiffer
Mica	Improves electrical resistance
Graphite	Reduces friction
Wood flour	Increases tensile strength but reduces water resistance

overall cost of the plastic. Often up to 80% of a plastic may be filler.

One form of additive used is a gas. The result is foamed or 'expanded' plastics. Expanded polystyrene is used as a lightweight packing material. Polyurethanes in the expanded form are used as the filling for upholstery and as sponges.

COMMON PLASTICS   *Polyethylene,* commonly known as *polythene,* is made in two forms (*Figure 5.5*) low-density or high-density. Low-density polythene is essentially a linear chain polymer with a small number of branches. The effect of this is that only limited crystallisation is possible. This results in a lower density than if complete crystallisation had been possible, hence the term 'low-density' applied to this form of polythene. High-density polythene is a completely linear polythene

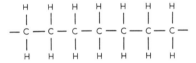

The simple polyethylene molecule

The linear chain which is easy to pack in an orderly array, i.e. a crystalline form

(a) High density polyethylene

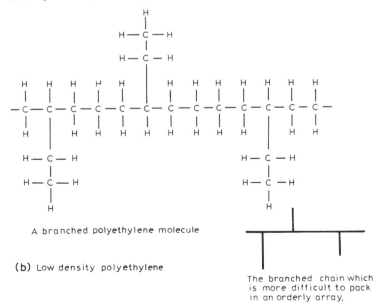

A branched polyethylene molecule

(b) Low density polyethylene

The branched chain which is more difficult to pack in an orderly array, ie. a crystalline form

Figure 5.5 Polyethylene molecules, (a) The basic polyethylene molecule, (b) A branched polyethylene molecule

and a high degree of crystallisation is possible, resulting in the higher density.

The following table gives a comparison of the typical properties of these two forms of polythene.

Property	Low density	High density
Density/10^3 kg m^{-3}	0.92	0.95
Melting point/°C	115	135
Tensile strength/N mm^{-2}	8–16	22–38
Elongation %	100–600	50–800
Maximum service temp./°C	85	125

Low-density polythene softens in boiling water, the high-density does not. Both forms are thermoplastics. Low-density polythene is used mainly in the form of films and sheeting, e.g. polythene bags, 'squeeze' bottles, ball-point pen tubing, wire and cable insulation. High-density polythene is used for piping, toys, filaments for fabrics and household ware. Both forms of polythene have excellent chemical resistance, low moisture absorption and high electrical resistance.

Low- and high-density polythene can be blended to give a material with properties between those of the two separate forms. The additives commonly used with polythene are carbon black as a stabiliser, pigments to give coloured forms, glass fibres to give increased strength and butyl rubber to prevent inservice cracking.

Polyvinyl chloride, commonly known as p.v.c., is a linear chain polymer with bulky side groups and so gives a mainly amorphous material (*Figure 5.6*). When used without a plasticiser it is a rigid,

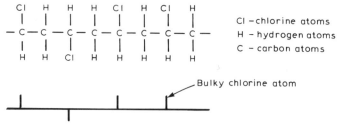

Figure 5.6   The basic form of a p.v.c. molecule. The chlorine atoms are generally arranged irregularly on the different sides of the chain, so rendering orderly packing of the chains difficult

relatively hard material. Most p.v.c. products are made with a plasticiser incorporated with the polymer. The amount of plasticiser is chosen to give a plastic with the required degree of flexibility, the amount varying between about 5 and 50%. Fillers, stabilisers and pigments are also often added.

The following table shows how the amount of plasticiser affects the properties of the p.v.c.

Property	No plasticiser content	Low plasticiser content	High plasticiser content
Density/10^3 kg m^{-3}	1.4	1.3	1.2
Tensile strength/N mm^{-2}	52–58	28–42	14–21
Elongation %	2–40	200–250	350–450
Maximum service temp./°C	70	60–100	60–100

The rigid form of p.v.c. is used widely for piping, but not for hot water, as it has a maximum service temperature of only 70°C. Above that temperature it softens too much. Plasticised p.v.c. is used for the fabric of 'plastic' raincoats, bottles, shoe soles, garden hose piping, gaskets and inflatable toys. All forms of p.v.c. have good chemical resistance, though not as good as polythene. It is a thermoplastic.

Polystyrene has bulky side groups attached irregularly to the polymer chain and so gives an amorphous structure, as it is not possible to pack such chains together in an orderly manner. A form can be produced which has the side groups arranged more uniformly to give a crystalline structure; it is however not generally used commercially. Polystyrene with no additives is a brittle, transparent, material with a maximum service temperature of only about 65°C (the crystalline form has much better temperature properties with a melting point of 273°C). This type of polystyrene finds its main uses as containers for cosmetics, light fittings, toys and boxes.

A toughened polystyrene can be produced by blending polystyrene with rubber particles. This gives a marked improvement in properties, the material being much less brittle. This material has a considerable number of uses, e.g. cups used in vending machines, casings for cameras, projectors, radios, television sets and vacuum cleaners.

Another very useful material can be produced by forming the polymer chains with three different polymer materials, polystyrene, acrylonitrile and butadiene. The product is called *acrylonitrile – butadiene – styrene terpolymer,* or just *ABS.* This material is tough, stiff and resists abrasion. It is used as the casing for telephones, vacuum cleaners, hair driers, radios, typewriters, etc, as well as for safety helmets, luggage, boat shells and food containers.

The following table shows typical properties of these various forms of polystyrene.

Property	Polystrene	Toughened polystyrene	ABS
Density/10^3 kg m^{-3}	1.1	1.1	1.1
Tensile strength/N mm^{-2}	35–60	17–42	17–58
Elongation %	1–3	8–50	10–140
Maximum service temp./°C	65	75	110

The above are all thermoplastic, amorphous materials. Polystyrene has good chemical resistance though can be attacked by some cleaning fluids. ABS has better chemical resistance.

Another widely used form of polystyrene is *expanded polystyrene,* which finds use as a rigid air-filled structure for insulation and packaging.

Polyamides, or *nylons* as they are more commonly called, are linear polymers and give rise to crystalline structures. There are a number of common polyamides: nylon 6, nylon 6.6, nylon 6.10 and nylon 11. The numbers refer to the number of carbon atoms in each of the substances reacted together to give the polymer. The full stops separating the two numbers are sometimes omitted, e.g. nylon 66 is nylon 6.6. Nylon materials are strong, tough, and abrasion resistant. They are thermoplastic materials with a relatively high softening temperature. Nylons tend to absorb moisture, this reducing the strength. Though they have reasonable chemical resistance there are some chemicals that attack nylon.

The following table shows the properties that are obtained with common nylons.

Property	Nylon 6	Nylon 6.6	Nylon 6.10	Nylon 11
Density/10^3 kg m^{-3}	1.1	1.1	1.1	1.1
Tensile strength/N mm^{-2}	70–90	80	60	50
Elongation %	60–300	60–300	85–230	70–300
Maximum service temp./°C	120	120	120	120

The effect of water absorbed by the nylon can be a reduction in the tensile strength of about 30 to 50%.

Nylons often contain additives; a stabiliser may be used for a nylon which is exposed to ultraviolet radiation. Flame retardant additives may be used for nylon components exposed to fire risk. Glass spheres or glass fibres may be added to give improved strength and rigidity. The following table shows the effects on nylon 6 of such additives:

Property	50% glass spheres	30% glass fibres
Tensile strength/N mm^{-2}	75	155
Elongation %	4	6

The glass-fibre-filled nylon has also much better elevated temperature properties, a maximum service temperature of 180°C being possible.

Nylon is used for the manufacture of fibres for use in clothing. Gears and bearings are made from nylon, with its low frictional and self-lubricating properties. Molybdenum disulphide is used as an additive with Nylon 6 to give a material with very low frictional properties. Glass-sphere or glass-fibre-filled nylon is used for the housings of power tools, formers for electrical coils, electric plugs and sockets. Nylon is used for motor car door handles, lock components, fans, bushes and bearings.

THE GENERAL PROPERTIES OF PLASTICS

Compared with metals, plastics have a low density, of the order of 0.9 to 1.4×10^3 kg m^{-3}. Filled plastics may have higher densities, while expanded plastics will have lower densities. Unlike metals, they have low thermal and electrical conductivities. They are thus widely used where electrical insulators are required, e.g. the casing of electrical plugs or the sleeving for electrical wires. Plastics are considerably less stiff and have lower tensile strengths than metals. Fibre-filled plastics can however have high strengths and stiffnesses. Plastics have a low hardness, being much more easily indented than metals. The mechanical properties of plastics deteriorate rapidly with an increase in temperature, many plastics being of no use at temperatures above about 100°C. Some plastics are transparent, but coloured forms can be produced by the addition of pigments. Plastics are reasonably resistant to water and acids but can often be damaged by organic solvents, the reverse of the resistance of metals which are prone to damage by acids but not usually by organic solvents.

The cost per unit mass of plastics, with no additives, is higher than that of some metals. On the basis of cost per unit volume they are comparable with that of metals. One of the important cost factors is, however, the much lower manufacturing costs associated with plastics. The following relative costs are based on that of unit mass of aluminium so that comparisons can be made with the table given on page 63.

Material	Relative cost per unit mass	Relative cost per unit volume
Aluminium	1	0.4
Polythene, low-density	0.4	0.4
Polythene, high-density	0.5	0.5
PVC	0.5	0.4
Polystyrene	0.5	0.5
ABS	1	1
Nylon	2	2

The following is an example of the information given by a manufacturer of plastics. (Courtesy of Bayer (UK) Ltd.)

Novodur (ABS polymer)

Appearance	Opaque, full gloss surface
Basic components	Acrylonitrile, butadiene, styrene
Physical structure	Amorphous
Density	1.03–1.06 Mg/m³ compact
	0.3–1.0 Mg/m³ expanded
Form supplied	Cubic granules
Bulk weight	500–600 g/l
Colour range	Available in natural colour and in all common opaque colours
Temperature performance	Maximum permanent service temperature without load: 80–105°C, according to type
Processing methods	Processing the raw material: Injection moulding, extrusion, compression moulding, extrusion blow moulding, foam moulding and extruding. Processing the semi-fabricated form: Thermoforming, e.g. vacuum forming, cold forming, e.g. deep drawing Machining: Sawing, drilling, turning, milling, tapping, die-cutting Jointing: Non-detachable: cementing, welding, nailing, riveting Detachable: clamping and snap fitting, screwing Decorating: Painting, printing, metallising, embossing, polishing
Predominant applications	Domestic appliances, the automotive sector, radios, television and audio equipment, furniture, office equipment, still and ciné photography, electrical goods, toys, equipment for leisure activities, and the textile industry

ELASTOMERS

Natural rubber is obtained from the sap of a particular type of tree. Both natural and synthetic rubbers are combined with additives such as plasticisers, anti-oxidants and fillers to give the required product. The products consist of linear chain molecules with some cross-linking between chains (*Figure 5.7*). This cross-linking is needed to ensure that the material will be elastic, i.e. return to its original dimensions when the load is removed. However, if there is too much cross-linking the material becomes less flexible and possibly even rigid. The introduction of sulphur into rubber to produce cross-links is called *vulcanisation*. Fully vulcanised rubber is the material known as *ebonite*.

The following table gives some of the properties of typical rubbers.

Before stretching (a)	When stretched (b)

Figure 5.7 Stretching an elastomer which has cross-links

Material	Tensile strength /N mm⁻²	Elongation /%	Service temp. range/°C
Natural rubber	30	800	−50 to +80
Butyl rubber	20	900	−50 to +100
Neoprene	25	1000	−50 to +100
Nitrile	28	700	−50 to +120
Silicone rubber	6	250	−80 to +235

Natural rubber and butyl rubber have relatively poor resistance to oils and greases. Silicone rubber has better resistance but neoprene and nitrile have very good resistance.

PROBLEMS

1. Distinguish between thermoplastic, thermosetting and elastomer materials on the basis of their elastic behaviour.

2. How does the behaviour of thermoplastic and thermosetting materials differ when they are heated?

3. Describe the difference between amorphous and crystalline polymer structures and explain how the amount of crystallinity affects the mechanical properties of the polymer.

4. Compare the properties of low- and high-density polythene and explain the differences in terms of structural differences between the two forms.

5. Why are (a) stabilisers, (b) plasticisers and (c) fillers added to polymers?

6. Describe how the properties of p.v.c. depend on the amount of plasticiser present in the plastic?

7. Which form of p.v.c. (i.e. without plasticiser, with a small amount or with a high amount) would be most likely to be used for the following applications.

  (a) Drain pipes.
  (b) Garden hose piping.
  (c) An inflatable toy.
  (d) A drinks bottle.

8. What is the effect on the properties of nylon of adding glass fibres?

9. How do the properties of plastics compare with those of metals?

10. Increasing the amount of sulphur in a rubber increases the amount of cross-linking between the molecular chains. How does this change the properties of the rubber?

11. A domestic bucket is to be made from polythene. Should high or low density polythene be used? Explain your answer.

12. Explain how elastomers can be stretched to several times their length and still be elastic and return to their original length.